Contents at a Glance

Introduction 1

Part 1: Get Going! 5
CHAPTER 1: Buying a Computer 7
CHAPTER 2: Setting Up Your Computer 33
CHAPTER 3: Buying and Setting Up a Printer 61

Part 2: Getting Up to Speed with Windows 75
CHAPTER 4: Working with Apps in Windows 77
CHAPTER 5: Six Great Apps that Come with Windows 107
CHAPTER 6: Managing Your Personal Files 127
CHAPTER 7: Making Windows Your Own 153

Part 3: Going Online 175
CHAPTER 8: Getting Connected to the Internet 177
CHAPTER 9: Browsing the Web 191
CHAPTER 10: Staying Safe While Online 213
CHAPTER 11: Keeping in Touch with Mail 233
CHAPTER 12: Working in the Cloud 253
CHAPTER 13: Connecting with People Online 269

Part 4: Having Fun 289
CHAPTER 14: Let's Play a Game! 291
CHAPTER 15: Creating and Viewing Digital Photos and Videos 305
CHAPTER 16: Listening to Music on Your PC 325

Part 5: Windows Toolkit 341
CHAPTER 17: Working with Networks 343
CHAPTER 18: Protecting Windows 361
CHAPTER 19: Maintaining Windows 373

Index 389

Table of Contents

INTRODUCTION. 1
About This Book 1
Foolish Assumptions 1
Icons Used in This Book 2
Beyond the Book 3
Where to Go from Here. 3

PART 1: GET GOING! 5

CHAPTER 1: **Buying a Computer** 7
What Can You Can Do with a Computer? 8
Get Up to Speed on Hardware 11
Input Devices: Putting Stuff In 15
Output Devices: Getting Stuff Out 16
What Is Software? 17
Decide on a Type of Computer 19
Choose a Display Type 24
Evaluate Your Storage Options 26
Consider How You Will Get Online 28
Where to Shop for Your New PC 30

CHAPTER 2: **Setting Up Your Computer** 33
Out of the Box: Set Up Your New PC 34
Set Up Windows 38
Take a First Look at Windows 40
Use a Mouse, Trackball, or Touchscreen 42
Get Familiar with the Start Menu 44
Sign Out and In 48
Switch Accounts 49
Lock Windows while You're Away 51
Restart Windows 51
Place the Computer in Sleep Mode 52
Shut Down Your Computer 53
Create Additional User Accounts 54
Change an Account's Type 57

Manage Family Settings 58

CHAPTER 3: **Buying and Setting Up a Printer** 61
Do You Need a Printer? 62
Choose the Right Printer 62
Unpack and Install a New Printer 64
Set Up a Printer to Work with Windows 64
Set a Default Printer 67
Set Printer Preferences 69
Manage a Print Queue 72
Remove a Printer 73

PART 2: GETTING UP TO SPEED WITH WINDOWS 75

CHAPTER 4: **Working with Apps in Windows** 77
Learn the Names of Things 78
Start an App 81
Exit an App 85
Find Your Way Around in a Desktop App 86
Find Your Way Around in a Microsoft Store App 90
Work with a Window 91
Switch among Running Apps 95
Move and Copy Data between Apps 97
Install New Apps 101
Remove Apps 104

CHAPTER 5: **Six Great Apps that Come with Windows** 107
Do the Math with the Calculator App 108
Write Brilliant Documents with WordPad 110
Jot Quick Notes with Notepad 114
Set Alarms and Timers 116
Keep Up on the Weather 119
Save Time with Cortana 121
Explore Other Windows Apps 124

CHAPTER 6: **Managing Your Personal Files** 127
Understand How Windows Organizes Data 128
Explore the File Explorer Interface 132
Move between Different Locations 134

Locate Files and Folders 136
View File Listings in Different Ways. 139
Select Multiple Items at Once. 142
Move or Copy an Item. 142
Delete or Undelete an Item 144
Rename an Item. 145
Create a Shortcut to an Item 146
Create a Compressed File. 148
Customize the Quick Access List 149
Back Up Files to an External Drive. 150

CHAPTER 7: **Making Windows Your Own** 153
Customize the Windows 11 Start Menu 154
Customize the Windows 11 Taskbar 156
Customize the Windows 10 Start Menu 157
Customize the Windows 10 Taskbar 160
Customize the Screen Resolution and Scale 161
Apply a Desktop Theme 163
Change Desktop Background Image. 164
Change the Accent Color. 165
Manage Desktop Icons 167
Add Widgets to the Desktop. 168
Make Windows More Accessible 170

PART 3: GOING ONLINE. 175

CHAPTER 8: **Getting Connected to the Internet** 177
What Is the Internet?. 178
Explore Different Types of Internet Connections 180
Identify the Hardware Required 184
Set Up a Wi-Fi Internet Connection 187
Assess Your Software Situation 189

CHAPTER 9: **Browsing the Web** 191
Meet the Edge Browser. 192
Search the Web 196
Find Content on a Web Page 199
Pin a Tab 200
Create and Manage a Favorites List 200

Use Favorites 203

View Your Browsing History 204

Print a Web Page 205

Create Collections 207

Customize the New Tab Page and the Home Page 208

Adjust Microsoft Edge Settings 211

CHAPTER 10: **Staying Safe While Online** 213

Understand Technology Risks on the Internet 214

Download Files Safely 216

Use InPrivate Browsing 219

Use SmartScreen Filter and Block Unwanted Apps 220

Change Edge Privacy Settings 221

Understand Information Exposure 223

Keep Your Information Private 226

Spot Phishing Scams and Other Email Fraud 228

Create Strong Passwords 230

CHAPTER 11: **Keeping in Touch with Mail** 233

Sign Up for an Email Account 234

Set Up Accounts in the Mail App 236

Get to Know the Mail Interface 238

Receive Messages 239

Reply to or Forward a Message 241

Create and Send Email 243

Manage Addresses 245

Send an Attachment 247

Change Mail Account Settings 248

CHAPTER 12: **Working in the Cloud** 253

Understand Cloud-Based Applications 254

Use Office Online 256

Access Your OneDrive Storage 257

Add Files to OneDrive 259

Share a Folder or File Using OneDrive 261

Create a New OneDrive Folder 263

Use the Personal Vault 264

Adjust OneDrive Settings 265

Configure Online Synchronization 266

CHAPTER 13: **Connecting with People Online** 269

Use Discussion Boards and Blogs 270
Participate in a Chat. 273
Understand Instant Messages 275
Explore Microsoft Teams 277
Explore Skype 280
Use a Webcam 282
Get an Overview of Collaborative and Social
Networking Sites 283
Sign Up for a Social Networking Service 285
Understand How Online Dating Works. 287

PART 4: HAVING FUN 289

CHAPTER 14: **Let's Play a Game!** 291

Learn the Types of Game Delivery. .291
Explore the Various Gaming Genres .293
Understand How Game Makers Get Paid. 299
A Few of My Favorites 300

CHAPTER 15: **Creating and Viewing Digital Photos and Videos** 305

Capture Pictures and Video with the Camera App 306
Make Audio Recordings with Voice Recorder 311
Find and Play Videos Using the Movies & TV App 314
Transfer Photos and Videos from a Camera or Phone. 318
View and Edit Photos in the Photos App. 319
Create a Video with the Video Editor. 321

CHAPTER 16: **Listening to Music on Your PC** 325

Preparing to Listen to Digital Music. 326
Introducing Windows Media Player. 330
Make Your Stored Music Available in Windows
Media Player 332
Play Music 333
Create a Playlist 335
Rip a Music CD 337
Burn a Music CD. 338
Acquire New Music 339

PART 5: WINDOWS TOOLKIT 341

CHAPTER 17: **Working with Networks** 343
Plan and Set Up a Home Network 344
Enabling Wireless Router Security 346
Set Up File Sharing on Your PC 348
Choose What Folders to Share 351
Share a Local Printer 354
Connect Bluetooth Devices to Your PC 357
Use Your Cell Phone as a Hotspot 358

CHAPTER 18: **Protecting Windows** 361
Choose Security Software 362
Update Windows 364
Check Windows Security Settings 367
Change Your Microsoft Account Password 369
Change How You Sign Into Windows 371

CHAPTER 19: **Maintaining Windows** 373
Shut Down an Unresponsive Application 374
Troubleshoot Application Problems 376
Repair or Remove an App 377
Set an App to Run in Compatibility Mode 378
Restore Your System Files 379
Use Windows Troubleshooter Utilities 382
Reset Your PC: The Last Resort 383
Free Up Disk Space 385

INDEX 389

Introduction

Computers for consumers have come a long way in just 40 years or so. They're now at the heart of the way many people communicate, shop, and learn. They provide useful tools for tracking information, organizing finances, and being creative.

During the rapid growth of the personal computer, you might have been too busy to jump in and learn the ropes, but you now realize how useful and fun working with a computer can be. In fact, for seniors, the computer opens up a world of activities and contacts they never had before.

This book can help you get going with computers quickly and painlessly.

About This Book

This book is specifically written for mature people like you — folks who are relatively new to using a computer and want to discover the basics of buying a computer, working with software, and getting on the Internet. In writing this book, I've tried to take into account the types of activities that might interest a 55-plus-year-old who's discovering the full potential of computers for the first time.

Foolish Assumptions

This book is organized by sets of tasks. These tasks start from the very beginning, assuming you know little about computers, and guiding you through the most basic steps in easy-to-understand language. Because I assume you're new to computers, the book provides explanations or definitions of technical terms to help you out.

All computers are run by software called an *operating system*, such as Windows, macOS, or Linux. Because Microsoft Windows–based personal computers (PCs) are the most common type of computer,

I am assuming that you have a desktop or laptop PC running Windows. Windows 11 is the latest version, but many people still use Windows 10, so I cover both in this book. There are also various types of mobile computers available, such as tablets and smartphones, but this book doesn't cover them.

Windows is updated periodically, and each time there are minor feature changes. This book is based on the late 2021 versions of Windows 10 and Windows 11. You may encounter minor differences between what you see onscreen and what you read about in this book, but they should not prevent you from getting things done.

Icons Used in This Book

Sometimes, the fastest way to go through a book is to look at the pictures — in this case, icons that draw your attention to specific types of useful information. I use these icons in this book:

WARNING

The Warning icon points to something that can prevent or cause problems.

REMEMBER

The Remember icon points out helpful information. (Everything in this book is helpful, but this stuff is even *more* helpful.)

TIP

The Tip icon points out a hint or trick for saving time and effort or something that makes Outlook easier to understand.

TECHNICAL
STUFF

The Technical Stuff icon marks background information you can skip, although it may make good conversation at a really dull party.

Beyond the Book

In addition to the material in the print or e-book you're reading right now, this product also comes with some access-anywhere goodies on the web. Check out the free Cheat Sheet for a checklist for buying a computer, computer care and maintenance tips, and Windows keystroke shortcuts. To get this Cheat Sheet, simply go to www.dummies.com and type **Computers For Seniors For Dummies Cheat Sheet** in the Search box.

Where to Go from Here

Whether you need to start from square one and buy yourself a computer or you're ready to just start enjoying the tools and toys your current computer makes available, it's time to get going, get online, and get computer savvy.

1

Get Going!

IN THIS PART . . .

Buying a computer

Finding your way around Windows

Buying and setting up a printer

IN THIS CHAPTER

» Seeing what you can do with computers

» Getting an overview of hardware

» Understanding the roles of input and output devices

» Appreciating operating systems and applications

» Deciding on a personal computer type

» Comparing display types

» Evaluating your storage options

» Considering your Internet options

» Shopping for your ideal PC

Chapter **1**

Buying a Computer

I f you've never owned a computer and now face purchasing one for the first time, deciding what to get can be a somewhat daunting experience. There are lots of technical terms to figure out and various pieces of **hardware** (the physical pieces of your computer such as the monitor and keyboard) and **software** (the brains of the computer that help you create documents and play games, for example) that you need to understand.

In this chapter, I introduce you to the world of activities your new computer makes available to you, and I provide the information you need to choose just the right computer for you.

Remember as you read through this chapter that figuring out what you want to do with your computer is an important step in determining which computer you should buy. You have to consider how much money you want to spend, how you'll connect your computer to the Internet, and how much power and performance you need from your computer.

What Can You Can Do with a Computer?

Perhaps your friends and family have been telling you that you need a computer, but have they explained why? Today's computers can do some pretty amazing things. Not only can they connect you to the wide world of the Internet, but they can run applications that let you store and organize photos, write your memoirs, make your own greeting cards, play all kinds of games, track your investments, and so much more.

The following list walks you through some of the things a computer will enable you to do. Depending on what activities are important to you, you can make a more-informed purchasing choice.

» **Keep in touch with friends and family.** The Internet makes it possible to communicate with other people via email; share video images using built-in video recorders or webcams (tiny video cameras that capture and send your image to another computer); and make phone and video calls using your computer and Internet connection to place calls with services such as Zoom and Skype. You can also chat with others by typing messages and sending them through your computer using a technology called **instant messaging** (IM). These messages are exchanged in real time, so that you and your grandchild, for example, can see and reply to text or share images immediately. Part 3 of this book explains these topics in more detail.

» **Research any topic from the comfort of your home.** Online, you can find many reputable websites that help you get information on anything from expert medical advice to the best travel deals. You can read news from around the corner or around the world. You can visit government websites to get information about your taxes and Social Security benefits, and go to entertainment sites to look up your local television listings or movie reviews.

» **Create greeting cards, letters, or home inventories.** Whether you're organizing your holiday card list, tracking sales for your home business, or figuring out a monthly budget, computer programs can help. For example, Figure 1-1 shows a graph that the Excel application created from data in a spreadsheet.

» **Pursue hobbies such as genealogy or sports.** You can research your favorite team online or connect with people who have the same interests. The online world is full of special-interest discussion groups where you can talk about a wide variety of topics with others.

» **Play interactive games with others over the Internet.** You can play everything from shuffleboard to poker and even participate in action games in virtual worlds. Love backgammon? Got you covered. Online bridge league? There are hundreds. Any game that you love offline, you can play online. You can play games with the computer, with total strangers, or (my favorite) with family and friends.

» **Share and create photos, drawings, and videos.** If you have a digital camera or smartphone, you can transfer photos to your computer (called *uploading*) or copy photos off the Internet (if their copyright permits it) and share them in emails or use them to create your own greeting cards. If you're artistically inclined, you can create digital drawings. Many popular websites make sharing your homemade videos easy, too. If you have a digital video camera or smartphone and editing software, you can use editing tools to make a movie and share it with others via video-sharing sites such as YouTube or by email. Steven Spielberg, look out!

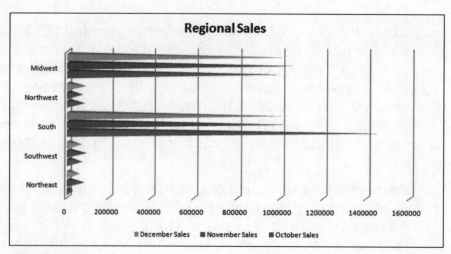

FIGURE 1-1

» **Shop online and compare products easily, day or night.** You can shop for anything from a garden shed to travel deals or a new camera. Using handy shopping site features, you can easily compare prices from several stores or read customer product reviews. Many websites, such as `pricegrabber.com`, list product prices from a variety of vendors on one web page, so you can find the best deals. Beyond the convenience, all this information can help you save money.

» **Manage your financial life.** You can do your banking or investing online and get up-to-the-minute data about your bank account, credit card balances, and investments. For example, Figure 1-2 shows Quicken, an application that enables you to track and view all your bank accounts and investments in one place.

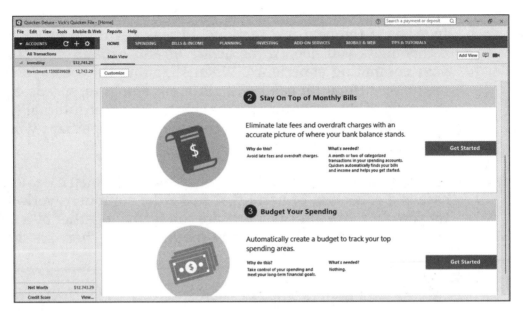

FIGURE 1-2

Get Up to Speed on Hardware

Your computing experience consists of interactions with hardware and software. I will explain both of those thing, but let's start with hardware. The *hardware* is all the tangible computer equipment — the parts you can see and touch.

REMEMBER

You should know a little something about computer hardware before you buy your first computer because the various components are available in a variety of quality and performance levels, and the component choices affect both a computer's price and its suitability for certain tasks.

WARNING

It's not always a good value to buy a top-of-the-line computer. In fact, unless you want to do something really specific and high-end, like professional graphic arts or movie production, high-end computers aren't usually worth the price. If all you want to do is write letters, share photos, and surf the Internet, don't waste your money. Get a moderately priced model that has the components you need.

In this and the next few sections, I break down the major hardware components you need to be aware of and explain how they affect your computing experience. When shopping for a computer, you'll see many different models with different amounts, speeds, and quality levels of the essential internal parts. The best internal parts cost more money, but offer better performance and perhaps will become obsolete less quickly.

All computers consist of some type of metal-framed case. Within the case is a collection of parts that make the computer work its magic for you. All those internal parts are connected together by a large circuit board inside the case called a **motherboard**. There are three main functions of the internal components:

TIP

» **Processing:** The computer's ability to receive input, perform an operation on it, and deliver output. The component in charge of processing is the **central processing unit (CPU)**. It's a small, very high-tech semiconductor chip mounted on the motherboard. Some people call the entire computer case the CPU, but that's not accurate.

CPU speed is rated in billions of hertz (**gigahertz**). The higher the GHz, the faster the processor. Generally speaking, the faster your CPU, the faster most applications run. That's not always true, though, because other components can cause bottlenecks, like slow Internet speeds, that can make a computer seem slow when its processor is just fine. The speed and features of the CPU make a big difference in the price of the computer.

There are lots of highly technical features that distinguish one CPU from another, but you don't need to worry about those for the most part; they're mostly of interest to people who are a lot geekier than you. For a basic home computer any of the CPUs available in new PCs today will be fine.

» **Memory:** The computer's ability to juggle the digital data that is active while the computer is running. Another name for memory is **random access memory (RAM)**. Its capacity is measured in billions of bytes (**gigabytes**). A **byte** is a group of eight binary digits (**bits**).

The more RAM a computer has, the more things it can do simultaneously. For example, a computer with a lot of RAM can run several complicated applications at once without bogging down. More RAM is better, but can also drive up the price of the system. The absolute minimum amount of RAM I would consider in a new PC is 8 gigabytes (GB). By the time you read this, though, the minimum might be higher — perhaps 16 GB.

» **Storage:** The computer's ability to keep a lot of data and many applications permanently on hand, ready to be copied into memory and used at a moment's notice. Every computer has at least one permanent storage component inside its case, known generically as a **hard drive** (or hard disk drive).

There's a PC for every budget and every set of needs — with prices to match. Before making your PC purchase, you need to think about how you'll use the computer and what specs it should have in order to enable you to do all the things you want to do with it.

Unless you are a big-time gamer who is super-serious about kicking butt in some graphics-intensive game, or you are a film-maker or a professional artist, any modern CPU should be fine for your needs. Don't spend a lot of money on the fastest CPU available. Most people won't notice the difference.

For the average home PC user, a far greater performance enhancer is the amount of RAM (memory) in the computer. The more RAM, the better. So if your budget allows you to have a fabulous CPU *or* a greater amount of RAM, definitely go with the RAM.

A hard drive's capacity is measured in gigabytes (billions of bytes) or terabytes (trillions of bytes). More capacity costs more. The more storage you have, the more applications can be installed on your computer, and the more data you can save (photos, documents, videos, and so on).

Hard drives can also use different technologies, which offer different data access speeds. Traditional hard drives (often referred to as **hard disk drives**, or HDDs) use a mechanical and magnetic storage system; they are slower and cost less for the same amount of capacity. **Solid-state drives** (SSDs) use storage media that is technologically similar to RAM, and they are faster, quieter, more reliable, cooler, and more expensive.

UNDERSTANDING HOW COMPUTERS STORE AND USE DATA

TECHNICAL STUFF

Computers store all data and programs digitally — in other words, using digits. At their most basic level, computers employ only two digits: 0 and 1. A number system with just those two digits is called a **binary** system. Each individual stored digit is called a **bit** (which is short for binary digit).

Computers work with binary data in a variety of different ways. For example, memory chips like RAM hold data using capacitors that can either store or not store a tiny electrical charge. A stored charge is a 1, and a lack of stored charge is a 0. Data is stored in memory using patterns of charge/no charge. This type of data storage is called **solid state** because there are no moving parts.

Solid-state storage can be either temporary (**dynamic**) or permanent (**static**). For example, the main RAM in a computer system is dynamic RAM, which is only temporary storage. The data stays in dynamic RAM only as long as the computer remains powered up. The instant it loses power, all the capacitors revert back to 0 (no charge). Dynamic RAM serves as a temporary work area for the computer whenever it is powered up. In contrast, solid-state drives store data in capacitors that don't lose their charge when the power goes out.

There are other ways that some older kinds of permanent computer storage holds its data. For example, hard disk drives store data magnetically, in patterns of positive and negative magnetic polarity. Optical discs such as CDs and DVDs store data in patterns of greater and lesser reflectivity on a shiny surface.

Input Devices: Putting Stuff In

Any computer system must provide at least one way for you to convey your wishes — in other words, to provide **input**. Almost all computers today offer at least two methods of accepting input: a keyboard and some type of pointing device.

REMEMBER

» A **keyboard** is similar to a typewriter keyboard. In addition to typing text, you can press certain key combinations to quickly issue commands for common activities such as selecting, copying, and pasting text.

All personal computers come with a keyboard. A desktop computer may have a detachable keyboard that you can replace with a different model if you want. There are some spiffy third-party keyboards you can get nowadays that have extra features and are designed to be more comfortable to use. A laptop keyboard is built-in, and you can't swap it out for a different one. (You can, however, connect a second, external keyboard to a laptop and use it instead of the built-in one.)

» **A pointing device** is a device that moves an on-screen **pointer** (usually shaped like an arrow). A pointer enables you to point at what you want and then select it by pressing a button on the pointing device. The three most popular types of pointing devices are:

- **Mouse**: A mouse is a little device about the size and shape of a bar of bath soap, with an LED and optical sensor on its underside. You slide the mouse across a flat surface with your hand, and that moves the pointer around onscreen. A mouse can be either wired (that is, have a cord that attaches to the computer) or wireless, operating via radio frequency (RF) signals.

- **Trackball**: A trackball is like an upside-down mouse. It has a stationary base with a ball on top, and you roll the ball with your fingers to move the onscreen pointer.

- **Touchpad**: On a laptop PC, a touch-sensitive rectangular pad in front of the keyboard serves as a pointing device. You move your finger across the touchpad to move the onscreen pointer, and tap the touchpad to select things. Figure 1-3 shows a touchpad.

FIGURE 1-3

REMEMBER

All computers come with either a mouse (on a desktop PC) or a touchpad (on a laptop). You can buy third-party pointing devices that you might like better than the default ones you get with your computer.

There are many specialty pointing devices available. Some monitors have touchscreen capabilities; if you have one of these, you can move your finger across the monitor screen to move the pointer or select things, like you might do on a touchpad.

Output Devices: Getting Stuff Out

It would be a pretty one-sided and unsatisfying experience if you never got anything back from your computer, right? The most common way a computer provides feedback is through the **display** screen.

The **display** is the graphical panel where you see the operating system interface, the applications you run, the websites you visit, and the data files you create, such as documents, spreadsheets, and messages. When a display is a separate unit from the computer, it's often referred to as a **monitor**.

A **printer** is another popular output device. A printer turns onscreen data to a paper copy that you can share with others. Chapter 6 covers printers in detail.

Yet another output device is a **speaker** (or a set of speakers). A computer speaker works basically the same way the speaker on your stereo system works — it enables you to hear the sound effects your computer generates as it operates, such as dings and beeps that accompany error messages. It also enables you to listen to music and watch videos that include sound using your computer. Chapter 18 explains how to play music on your computer.

What Is Software?

Software is what makes computer hardware work and lets you get things done, such as writing documents with Microsoft Word or playing a game of solitaire. A computer uses two types of software: operating systems and applications.

All computers have an **operating system (OS),** which is system software that starts up the computer and keeps it running as you use it. Examples for desktop and laptop computers include Microsoft Windows, macOS (for computers made by Apple), and Linux (a free operating system popular with techie-types). A lot of the upcoming chapters in this book explain how to interact with Microsoft Windows; I picked Windows to talk about in this book because it's the overwhelming favorite, with something like a 95 percent market share.

Mobile devices like tablets and smartphones have different operating systems. The most popular operating systems for mobile devices are iOS for Apple devices (iPhones and iPads) and Android for most other phones and tablets.

The operating system is responsible for the **graphical user interface (GUI),** which is pronounced *gooey*). The pictures, text, menus, boxes, and other items you see on the computer's screen are all part of the GUI. It also handles various housekeeping tasks like saving and opening files, talking to the hardware on your behalf, and starting and exiting applications.

An **application** (sometimes called an app or a program) is software that does something that's directly useful or beneficial to the human using the computer. For example, Microsoft Word is an application that helps you write letters and other documents, and Microsoft Edge is a web browser application that helps you view web pages.

Each operating system comes with a few basic apps. For example, Microsoft Windows comes with a simple word processing program called WordPad, a simple drawing program called Paint, and a digital music player called Windows Media Player. Chapter 10 showcases several of these built-in Windows apps.

An operating system also comes with **utilities**, which are applications designed to perform tasks that keep your computer in top shape. For example, antivirus apps protect your computer from viruses (covered in Chapter 20) and Windows Update keeps Windows current.

TIP

You can actually do quite a bit with just the free apps that come with Windows (which I cover in Chapter 10), but if you ever want more, "more" is certainly available. For example, Microsoft Office is a suite of professional-quality business applications that you can subscribe to for about $100 a year, and use it on up to 5 computers. Chapter 8 covers applications in detail, and explains how to acquire, install, update, and remove them.

REMEMBER

Each app is written for one specific operating system; you can't mix and match them. However, most of the popular apps are available for multiple operating systems, so you just have to make sure you are are getting the version of the app that is for your OS.

WINDOWS VERSIONS AND EDITIONS

Each application has a version number or name, which is like a generation. Some version numbers correspond to the year the software was released, like Office 2021. Other version numbers indicate how many versions have come before, like Camtasia 13 (that is, there were 12 previous versions). Still others have nonnumeric names, like Adobe Acrobat CC.

Microsoft used to assign different version numbers or names to each generation of Windows (such as Windows XP, Windows 7, Windows Vista, and so on), but they stopped doing that for Windows back in 2015, when they released Windows 10. Windows 11 came out in 2021, and that's the current version at this writing. A feature called Windows Update runs automatically in the background, downloading and installing fixes and new features whenever they are available. Chapter 20 explains more about Windows Update.

Because not every computing situation has the same needs, Microsoft produces different *editions* of Windows with different subsets of features. The two main ones you will probably encounter are Home and Pro. As you might guess, the Home edition is for people who use their computers at home or in small businesses. It's a less expensive edition because it doesn't include some business-oriented networking and security features. All the instructions and advice in this book apply equally well to either of those editions.

Decide on a Type of Computer

A **personal computer (PC)** is a computer designed to be used by one person. The word "personal" helps distinguish PCs from a variety of other computer types, everything from the powerful servers that calculate trajectories at NASA to the computer chip that controls the temperature in your refrigerator.

There are two basic categories of PCs: those that are easily portable and those that aren't. So your first decision is: *how important is portability?* Will you be sitting at the same desk each time you use the computer, or will you be out-and-about in coffee shops, hotel rooms, and such? The more portable models are **laptops** (sometimes called **notebooks**), and the less portable ones are **desktops**.

Desktop PCs are big and rather heavy; you can't just throw one in a briefcase and hit the open road. But you tend to get more features and power for your money with a desktop, and they're easy to repair and upgrade. Desktops can have very large external monitors, which is good news if your eyesight isn't great and you need a big screen. And because the keyboard is detachable, you can use any keyboard that works well for you — such as an ergonomically designed model that's easy on the hands and wrists.

Laptops are lightweight and easy to handle. They can run on battery power for hours at a time, so you can use one anywhere — like in the passenger seat of a car, or in the middle of a national park. The built-in display, keyboard, and touchpad means there are no cables to get tangled or come unplugged, too. Because there is less space for fans, laptops sometimes run hot to the touch. Laptops tend to be more expensive for the same hardware capabilities, though, and more difficult (read: expensive) to repair. Most laptops have very limited upgrade possibilities.

WARNING

Tablets such as the iPad offer many basic computing capabilities and are extremely lightweight and portable. You can read books, check your email, play games, listen to music, watch videos, and more. A tablet can't take the place of a real PC, though. It doesn't have a real keyboard (although you can buy an add-on wireless keyboard if you really want one), so your finger on the touch-screen has to substitute for both keyboard and mouse. There's no expandability or upgradability, and most tablets don't even interface with common external devices like printers. Enjoy your tablet — or your mobile phone — but don't expect it to have all the same capabilities as a PC.

The traditional type of desktop PC is a tower design, which is a big rectangular box that sits upright on or under your desk. All the other components plug into it with cables: monitor, keyboard, mouse printer, and so on. Figure 1-4 shows an example.

A tower case has plenty of room to add upgrades and new capabilities. It has powerful inside fans that keep everything cool. It's easy to open the case and easy to remove and replace parts.

Courtesy of Dell, Inc.

FIGURE 1-4

Another form of desktop computer is an *all-in-one*, like the one shown in Figure 1-5. With this type of computer, there is no big separate case for the processing components. All the components that would normally be inside the tower case are instead compactly sandwiched into a smaller box built into the back of the monitor. All-in-ones are often sleek and modern looking, involve fewer cables, and save you from using floor space for a computer tower. However, repairing or customizing the hardware is awkward and difficult, and replacing parts can be expensive because they don't usually use generic, off-the-shelf parts.

Both desktop towers and all-in-ones take up more surface space than a laptop computer, but if you don't need portability in your computer or more space, a desktop may be the best choice.

FIGURE 1-5

Laptops come in a variety of sizes, mostly determined by the screen size (which is measured diagonally). The smallest mini laptops have screen sizes as small as 9"; the largest ones have screen sizes of 17" and upward. Larger displays are easier to see, but larger laptops are also bulkier and heavier (sometimes as much as 8 to 10 pounds), use more power, and can't run as long on a single battery charge. The smallest laptops may have undersized keyboards that are awkward to type on, and may have fewer features or less capability. For many people, a screen size of between 14" and 16" is a good compromise. Figure 1-6 shows a moderately sized model.

Another laptop feature to consider is a touchscreen. Many of today's laptops (and some desktops) have touchscreens that allow you to interact with them using your finger or a digital pen, like you would with a tablet computer. See Chapter 2 for advice on using a touchscreen computer.

TIP

If you can't decide between a small touchscreen laptop and a tablet computer, you might consider a two-in-one laptop (see Figure 1-7), which enables you to either rotate the monitor to rest on the back of the keyboard or remove the monitor so you can use the laptop like a tablet. When you have no active physical keyboard, you have to use the touchscreen feature to interact with the laptop.

FIGURE 1-6

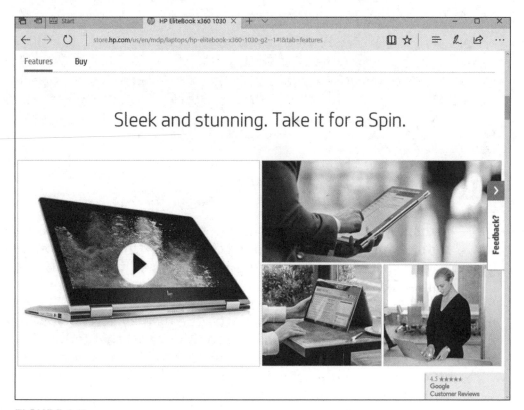

FIGURE 1-7

WHAT PORTS SHOULD A COMPUTER HAVE?

Each desktop or laptop PC has one or more **ports** (connectors) to enable you to plug in external peripherals, like printers, speakers, scanners, and so on. A **peripheral** is a computer component that isn't built into the main case. Desktop PCs tend to offer more ports than laptops do.

Universal serial bus (USB) is a very common, generic port type used for many different kinds of devices. Nearly every external device type comes in a USB version.

Having more USB ports on your computer is nearly always better. The more USB ports your computer has, the more peripherals it can support at once. That might not seem like a big deal . . . until that moment when you need to plug in an external keyboard *and* trackball *and* printer *and* webcam all at once.

Let's pause a moment on that scenario of needing to plug in more USB peripherals at once than your computer has ports to accommodate. One way around that is to connect a USB hub (which is kind of like a power strip, but for USB instead of electricity) to a USB port. Then you can connect multiple USB devices to the hub and run them all off of a single USB port on the computer. It's not ideal, but it's a workable workaround.

Choose a Display Type

A display screen is the window to your computer's contents. The right display can make your computing time easier on your eyes. The crisper the image, the more impressive your vacation photos or that video of your last golf game will be.

REMEMBER

The display consists of two parts: the display screen you look at, and the display adapter (a.k.a. graphics card) that tells the screen what to display. On a laptop or an all-in-one desktop, you're stuck with the display screen and display adapter that it comes with; you can't usually upgrade them. That means you need to choose carefully when you purchase. Many online retailers enable you to choose the display adapter and display screen specifications when you place your order. On a tower desktop you can usually swap out both the display adapter and the monitor for better ones later.

The **display adapter** is the translator between the software and the display screen. It does the calculations that result in certain parts of the screen lighting up with certain colors. A high-end display adapter can add hundreds of dollars to the overall cost of a computer, so you don't necessarily want the fanciest and best one out there unless you plan on playing graphics-intensive games competitively or creating your own animated cartoon movies. The standard display adapter that comes with the average desktop or laptop PC today should be fine for ordinary home needs like email, word processing, and Internet exploration.

Some desktop computers come with a monitor; if the one you're interested in doesn't you'll need to buy a monitor separately. It is often actually better to buy the computer and the monitor separately because then you get exactly what you want.

Consider these factors when choosing a monitor:

» **Size:** Monitors for the average computer user come in all sizes, from tiny 9-inch screens on smaller laptops to 28-inch desktop models. Larger screens are typically more expensive. Although a larger monitor can take up more space side to side and top to bottom, many don't have a bigger **footprint** (that is, how much space their base takes up on your desk) than a smaller monitor.

» **Image quality:** The image quality can vary greatly between monitors. There are objective measurements of quality you can use to compare different models, but it's often better to simply go to a retail store that sells monitors and view the screens in action, like you do when shopping for a television. You can see at a glance which monitor's picture you prefer without having to fuss with technical specifications.

» **Resolution:** A monitor's resolution represents the number of tiny dots (called **pixels**) that form the images you see on the screen. The higher the resolution, the more pixels it contains and the crisper the image. The larger the screen size, the higher the resolution should be. For example, a 15" monitor should have at least a 1280 x 800 resolution; a 20" model should have at least 1680 x 1050.

» **Cost:** The least-expensive monitor might be the one that comes with your desktop computer, and many of these are perfectly adequate. You can often upgrade your monitor when you buy if you customize a system from a company such as Dell or Hewlett-Packard. Monitors purchased separately from a computer can range from around $100 to $2,000 or more. Check out monitors in person if possible to verify whether their image quality and size are worth the money.

» **Touchscreen technology:** Windows provides support for using a touchscreen interface, which enables you to use your fingers to provide input by tapping or swiping on the screen itself. If you opt for a touchscreen device, you can still use your keyboard and mouse to provide input, but touchscreen technology can add a wow factor when performing tasks such as using painting software or browsing around the web or an electronic book (ebook).

WARNING

Displays with touchscreens usually cost more, and a lot of people who get them find that they don't use the touchscreen as much as they thought they would.

Evaluate Your Storage Options

Storage is an important part of a personal computer. You need to be able to store important software and data inside the computer case — in other words, internally. That's because internal storage is always available, and usually quite fast to access. You also need a way of getting access to external storage, such as external DVD and hard drives, USB flash drives, and online storage services.

All PCs have some sort of built-in storage device. Until recently, in nearly every system that was the **hard disk drive (HDD)**, sometimes just called the hard disk or hard drive. A hard disk drive is a sealed metal cartridge containing metal platters on which data is stored in patterns of magnetic polarity.

Today, a new technology of internal storage called a **solid-state drive (SSD)** is fast replacing HDDs as the storage medium of choice in PCs. SSDs use the same technology as RAM uses, and they are faster, more reliable, quieter, and cooler — in other words, all-around better. The one way that HDD wins that match-up is on price: SSD drives are more expensive for the same capacity.

Capacity is the other big decision point when computer-shopping. The higher capacity the internal storage device, the more applications and data you will be able to store without running out of room. Storage is measured in gigabytes (billions of bytes) or terabytes (trillions of bytes). More is better — to a point. When you get into very high-capacity storage devices (multiple terabytes), the price begins to get so much higher that it's not worth the cost for most casual users.

DO I NEED A DVD DRIVE?

In the recent past, most computers came with an optical drive where you could insert a DVD and play a movie or music. If you bought an application, it usually came on a DVD.

Today, many new computers and laptops don't include an optical drive for reading DVDs, partially because you can so easily stream video or download a new application from an online source without ever handling a DVD. People who still have software or videos on DVD that they need to be able to access using their computer can purchase an external DVD drive and connect it to the computer using a USB port whenever the drive is needed. And with most desktop tower PCs, it's a fairly simple matter to have an internal DVD drive installed at any local PC repair shop.

If you want to play the latest optical discs, get a computer with a Blu-ray player (or buy an external Blu-ray player that you can connect to your PC via USB port). Blu-ray is a great medium for storing and playing back feature-length movies. because it can store 50GB or more, about ten times as much as the average DVD.

WARNING

You also don't need to waste your money on an obscene amount of storage that you will never use. The average home user won't even need a single terabyte of storage in their lifetime. And keep in mind that you can have external storage as well as internal; at any point you can buy an external SSD or HDD and connect it to your computer to offload data that you might not need in the near future.

Consider How You Will Get Online

You will also have to decide how you'll connect to the Internet. You need Internet access because so much of the usefulness of a PC depends on being able to access online content.

From a computer-buying perspective, there is just one important consideration: how will your computer physically connect to your home Internet connection? Your two choices are wired (that is, with a cable) and wireless (with radio waves). A wired connection is faster and more reliable, but it keeps your computer tethered to one spot in the house. If you're going with a desktop PC, that's not a problem; in fact, most desktop PCs come with a port for plugging in a network cable. This port is commonly called an **Ethernet port**, or an RJ-45 jack, or wired networking.

Some desktop PCs also support wireless networking; if yours doesn't, it's easy enough to upgrade it to have that capability. Any local PC repair shop can install it.

If you've decided on a laptop, it had better have wireless networking capabilities built in, because you'll be roaming around the house with your laptop — and perhaps farther afield than that. There have been various standards for wireless networking (also called Wi-Fi) over the years, but all the standards start with 802.11 and have one or more letters following that number. The current standard at this writing is 802.11ax, also called WiFi 6; the slightly older standard of 802.11ac (WiFi 5) is also still in use. A new laptop should support one of these.

LET'S MAKE THIS SIMPLE . . .

If you're a bit overwhelmed by all this new vocabulary, never mind all that: just answer these questions.

Question 1: Desktop or laptop? Desktops are good for people who plan on using the PC from the same desk every day. You get more computing capability for your money with a desktop. Laptops offer more mobility and flexibility, but they have smaller screens and keyboards and are more difficult and costly to repair and upgrade.

Question 2: Do you need high performance? If you're going to be working extensively with video editing, music production, or the latest shoot-em-up games, look for a fast processor and a top-quality display adapter. If you're just interested in email, writing, and photo sharing, the latest processing and graphics technology will be wasted on you.

Question 3: Are you a multitasker? If you think you'll be doing a dozen things at once on your computer, like watching a video, editing photos, switching between different websites, and video chatting with a friend, you will need extra RAM (memory).

Question 4: Do you plan on using a lot of external devices? If you think you'll be plugging in many different devices, like an external keyboard, mouse, webcam, phone charger, speakers, and so on, you want a model with as many USB ports as possible.

Question 5: Wired or wireless Internet? A desktop PC should have an Ethernet port for connecting a cable that will link your computer to your Internet connection at home. A laptop PC should have wireless networking support (802.11ac or 802.11ax).

Internet service doesn't just magically appear at your home when you buy a computer; you have to subscribe to an Internet service through a company called an Internet Service Provider (ISP), which operates similarly to a cable TV company. (In fact, many cable TV companies also provide Internet service.) See Chapter 11 to learn how to choose and set up a home Internet connection.

Where to Shop for Your New PC

You can buy a PC for anywhere from about $199 to $5,000 or more, depending on your budget and computing needs. You might start shopping thinking that you want a "just the basics" model, but when you start thinking about extras such as a larger monitor or larger storage capacity, you may find that the price goes up quickly.

TIP

A good rule of thumb is to buy just as much computer as you need *now*, and don't plan too aggressively for what you might need in the future. By the time the future gets here, a new computer with better capabilities will probably be a lot cheaper than it is now.

You can shop in a retail store for a computer or shop online using a friend's computer (and perhaps get their help if you're brand new to using a computer). Consider researching different models and prices online with the help of a computer-savvy friend and using that information to get the best buy. Be aware, however, that most retail stores have a small selection compared to all you can find online on websites such as Amazon.com and Newegg.com. Additionally, retail stores sometimes carry slightly older models than those available online.

Buying a computer can be confusing, but here are some guidelines to help you find a computer at the price that's right for you:

>> **Determine how often you will use your computer.** If you'll be working on it eight hours a day running a home business, you will need a better-quality computer to withstand the use and provide good performance. If you turn on the computer once or twice a week just to check email, it doesn't have to be the priciest model in the shop.

>> **Consider the features that you need.** Do you want (or have room for) a 28-inch monitor on your desk? Is it critical that your computer runs very fast, or runs many programs at once? Do you need to store a great deal of data, such as hundreds of hours of video footage? (I cover computer speed and storage

later in this chapter.) Each feature or upgrade adds dollars to your computer's price. Understand what you need before you buy.

» **Shop wisely.** If you walk from store to store or do your shopping online, you'll find that the price for the same computer model can vary by hundreds of dollars at different stores. See if your memberships in organizations such as AAA, AARP, and Costco make you eligible for better deals. Consider shipping costs if you buy online, and keep in mind that many stores charge a restocking fee if you return a computer you aren't happy with. Some stores offer only a short time period, such as 14 days, in which you can return a computer.

» **Buying used or refurbished is an option, though new computers have reached such a low price point that this may not save you much.** In addition, technology gets out of date so quickly that you might be disappointed buying an older model, which might not support newer software or hardware.

» **Online auctions are a source of new or slightly used computers at a low price.** However, be sure you're dealing with a reputable store or person by checking reviews others have posted about them or contacting the online Better Business Bureau (www.bbb.org). Be careful not to pay by check (this gives a complete stranger your bank account number); instead use the auction site's tools to have a third party handle the money until the goods are delivered in the condition promised. Check the auction site for guidance on staying safe when buying auctioned goods.

TIP

Some websites allow you to compare several models of computers side by side, and others, such as Pricegrabber.com, allow you to compare prices on a particular model from multiple stores.

TIP

New to all this? Find a computer–savvy friend to help you shop for your first computer. People who have been using computers for awhile usually have an informed opinion about what features are important, what's a good value, and what pitfalls to avoid.

IN THIS CHAPTER

» Setting up a new computer

» Setting up Windows

» Using a mouse, trackball, or touchscreen

» Switching user accounts

» Locking Windows while you're away

» Signing in and signing out

» Restarting Windows

» Placing the computer in Sleep mode

» Shutting down the computer

» Creating additional user accounts

» Changing an account type

» Managing family settings

Chapter **2**

Setting Up Your Computer

When you unpack your new computer, you may need help getting it set up. That's what this chapter is for — to be your personal setup assistant! You'll learn how to connect up the components for a desktop or laptop (spoiler alert: a laptop is much easier!). I explain how to do the initial setup for Windows, and how to start up and shut down your new computer. You get some tips for using a mouse, trackball, or touchscreen, and I share some time-saving keyboard shortcuts.

Windows enables you to create multiple user accounts. Having a separate account for each person using the computer gives everyone some privacy by keeping personal files and folders separate. You can also create child accounts for any kids who may be using your computer, like visiting grandchildren. I explain how to create accounts, both the adult and the child kind, and how to switch between them.

Out of the Box: Set Up Your New PC

A desktop PC comes with a power cord, monitor, keyboard, and mouse — and maybe some other peripherals too, like speakers. You should connect all these components before turning on the computer.

TIP

Most desktop PCs come with a setup diagram or flyer that explains where to plug in each of the cords, but you can usually figure it out yourself just by comparing the plugs on the cables to the sockets on the PC. A laptop PC doesn't have much to connect — probably just a power cord.

A desktop PC comes with a standard three-prong power cord. Connect one end of it to the computer and the other to a wall outlet (or better yet, a surge-protected power strip).

A laptop PC comes with a power cord with a transformer block built into it — in other words, a big rectangular block. When you first unpack it, this power cord will probably be in two pieces. You need to connect the pieces to make it work.

If you have an external monitor, it will have two cables associated with it: a power cord and a cord that connects it to your computer. Over the years, there have been different monitor connector types, but today the standard is High Definition Multimedia Interface (HDMI), the same type of connector that modern TVs use to connect to cable boxes and DVD players. Figure 2-1 shows an HDMI connector. A built-in display screen doesn't require any special connection.

FIGURE 2-1

Your computer offers several types of connection ports (connectors accessible from the outside of the computer) that allow you to connect other devices, such as external mice and keyboards.

As I mentioned in Chapter 1, Universal Serial Bus (USB) is the most common type; nearly all wired peripherals (except most monitors) plug into USB ports, including nearly all external keyboards and pointing devices. Some external speakers and microphones also connect via USB, as well as scanners, printers, digital cameras, and chargers for mobile devices like smartphones and tablets.

Even wireless devices typically use USB ports. Yes, really! Wireless keyboards and mice come with a tiny wireless transmitter/receiver that plugs into a USB port. It then communicates wirelessly with the keyboard or mouse. Figure 2-2 shows a wireless mouse and its USB-based transmitter.

If your computer does not have enough USB ports to accommodate all the USB peripherals you want to connect, you can use a USB hub (shown in Figure 2-3), which allows you to plug extra peripheral devices into a single USB port on your computer.

WARNING

Sometimes high-speed USB devices don't work as well on hubs, so if you need a hub, try to put your low-bandwidth devices on it, like keyboard and mouse.

FIGURE 2-2

FIGURE 2-3

You don't have to connect your new computer to the Internet before you set it up; you can do that later. You might not even have Internet service at your house yet, and that's okay.

However, if you do happen to already have Internet available, you might as well use it. As Windows is finalizing its setup, it checks to see whether the Internet is available, and if it is, Windows downloads any updates it needs.

Most home Internet setups have a cable running from an external wall into the house, kind of like with cable TV. That incoming cable plugs into a box called a **broadband modem** that has some flashing lights on it. Your computer connects to the modem to get Internet service.

If you have a desktop PC, you probably have an Ethernet port on your computer. It's a port that accepts a networking cable like the one shown in Figure 2-4. Plug one end of the cable into that port on your computer, and the other end into the router. Boom, you're connected.

FIGURE 2-4

If you have a laptop PC, you'll want to connect to the router wirelessly for maximum mobility around the house. You can't do that until you get started setting up the computer, so put a pin in that for the moment.

Set Up Windows

With your computer set up, you're ready to turn it on. Start by pressing the power button on your computer to start up Windows.

The first time you start up a computer with Windows installed on it, you're prompted to walk through a setup procedure. If you're given the option, choose Express Settings to accept the default settings where possible. Some of the settings you'll be asked about include the following. Depending on the version and edition of Windows and how the PC manufacturer has set it up, these options may come in a different order than shown here, and you might not get asked all of them.

» **Language:** Choose the language you prefer, if asked. You will probably only be able to choose between UK and US English.

» **Region:** Choose the country where you are living. This makes a difference in things like currency symbols.

» **Keyboard layout:** Choose the keyboard layout that matches your keyboard language. It's probably US.) If you're given the option to add a second keyboard, skip that.

» **License agreement:** This is a long page of legalese. Read it if you want; then click Accept.

» **Internet:** If Windows setup automatically detects your Internet connection, it doesn't ask you about Internet; it just uses it. If you see something about Internet, it probably means you don't have Internet access yet. Without Internet access, you may need to create a Local account in order to get into Windows for now. Follow the prompts for that if you're asked.

» **Sign in with Microsoft:** You need a Microsoft account to take full advantage of Windows features. If you don't already have one, you'll be prompted to create one.

If you have Internet access, you'll be asked to sign in with your Microsoft account — which you might or might not already have. If you already have a Microsoft account, enter your email address for it, click Next, and then enter the password for it.

If you don't already have a Microsoft account, you'll need to create one. If you already have an email address, you can use it for your Microsoft account. The password you specify here is *not* the same as the password you already use to send and receive email; you should create a new, different password for your Microsoft account, even though it is using the same email address.

If you don't already have an email address, when Windows setup asks you to sign in with Microsoft, look for an option for creating a new email address, and then follow the prompts to do so.

» **Activity history:** If you are asked about sending Microsoft your activity history, click Yes or No as desired. Yes enables you to access your activity history across multiple devices when they are all signed in using the same Microsoft account; it's handy, but some people see it as a privacy invasion. If you only have this one computer, it's a moot point, so choose No.

» **Privacy settings:** Choose Yes or No for each of the privacy settings that Windows setup asks about. Then click Accept.

» **Windows Hello:** This is a feature that enables your computer to take advantage of facial recognition technology. Simply by sitting in front of your computer, you're signed in, in effect using your face as your password (provided your computer has a built-in camera — and many do these days, especially laptops). If you're prompted to set that up during Windows setup, you can follow the prompts to set it up.

» **Set up a PIN:** You can create a PIN to use in place of a password. If you are asked about this, follow the prompts to specify a PIN. You can set this up later in the Settings app in Windows if you want.

TIP

If you're connected to the Internet, you may see a prompt asking if you want to allow your PC to be discoverable by other PCs and devices on the network. Microsoft recommends choosing Yes for home and work networks and No for public networks like coffee shops and airports.

WARNING

Remember the email address and password you chose for your Microsoft account — you'll need them! If you chose a PIN for sign-in, make sure you remember that too. Security experts tell you not to write them down, but that poses its own challenge, especially if you are forgetful like I am! If you do write them down on paper, keep that paper somewhere secure and out of sight. That means *not* on a sticky note on your monitor!

Take a First Look at Windows

After you've worked through all the prompts, the Windows desktop (finally) appears. Chapter 3 covers Microsoft Windows in detail, so I don't want to steal its thunder here. Instead, I'll just give you these really attractive images of the Windows desktop to admire in Figures 2-5 and 2-6 and point out a few key activities and commands you need to know to work through the rest of this chapter. Figure 2-5 shows Windows 11, the latest version of Windows, and Figure 2-6 shows Windows 10, the next-to-latest version (still very popular). By comparing your screen to these figures, you can determine which version of Windows you have.

Here are the key features to note in Figures 2-5 and 2-6:

» The **desktop** is the background on which everything else sits. It has a default graphic, but you can set it to show any graphic or photo you like.

» The **taskbar** is the bar at the bottom of the screen. It shows buttons for each currently open app; you can click one to make that app's window active. (There aren't any open apps in Figures 2-5 and 2-6, so just put a pin in that knowledge for the moment.)

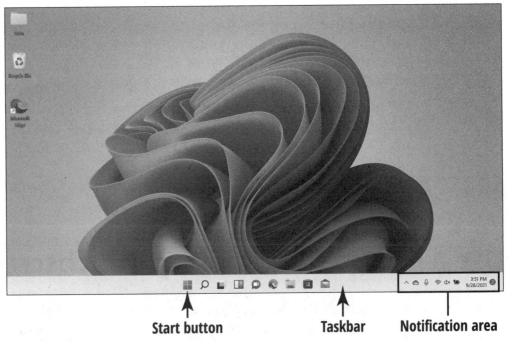

Start button   **Taskbar**   **Notification area**

FIGURE 2-5

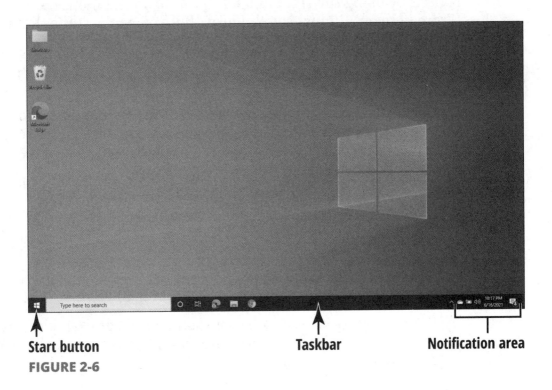

Start button   **Taskbar**   **Notification area**

FIGURE 2-6

» The **Start button,** which opens the Start menu, is your gateway to running applications. In Windows 11, it's the leftmost icon centered on the taskbar, as shown in Figure 2-5. In Windows 10, it's on the far left end of the taskbar (as in Figure 2-6).

» The **notification area** shows the current date and time, as well as shortcut icons (pictures) to utilities and services running in the background, such as the volume control for your speakers.

There are a lot more things going on in Figures 2-5 and 2-6, of course, but this list should serve for the moment.

Use a Mouse, Trackball, or Touchscreen

When you slide your mouse around on your desk (or roll a ball on top of a trackball), a corresponding pointer moves around your computer screen (a.k.a. the Windows desktop). You control the actions of that pointer by using the right and left buttons on the mouse or trackball. (From here on, I'll just say "mouse" but you can assume that I mean "or trackball" each time.)

Here are the main functions of a mouse and how to control them:

» **Click:** Move the onscreen pointer over the command or item you want to select and then press and release the left mouse button once.

Clicking has a variety of uses. You can click the Start button in Windows to open the Start menu, for example, and you can click a command on the Start menu to execute that command or open the app it represents.

» **Double-click:** Move the onscreen pointer over the item you want to activate and then press and release the left mouse button twice quickly in succession.

Double-clicking is used to activate something. For example, you can double-click an icon on your desktop to open the application or window it represents.

- » **Right-click:** Move the onscreen pointer over the command or item and then press and release the right mouse button once. Right-clicking makes Windows display a shortcut menu that's specific to the item you were pointing at. For example, if you right-click a picture, the menu that appears gives you options for working with the picture. If you right-click the Windows desktop, the menu that appears lets you choose commands that display a different view or change desktop properties.

- » **Drag:** Depending on the context in which it's used, dragging either moves (or copies or resizes) an item or selects data.

   To move something by dragging it, point at the item and then press and hold down the left mouse button as you move the pointer to a different location. For instance, you can drag an icon on the desktop to move it, or drag the title bar of a window to move the window. You can also move or copy files in File Explorer by dragging them. (We're getting ahead of ourselves here, though; file management is covered in Chapter 9.)

   To make a selection in a text editing program by dragging, click in the text where you want the selection to begin and then drag up, down, right, or left to highlight and extend the selection. Any action you perform after making a selection, such as pressing the Delete key on your keyboard or clicking a button for bold formatting, is performed on the selected text.

- » **Scroll:** Many mouse models have a wheel in the center that you can roll up or down to scroll through a document or website on your screen. Just roll the wheel down to move through pages going forward, or scroll up to move backward in your document. Some trackballs also have some sort of scrolling capability; my favorite trackball, the Kensington Slimblade, enables me to scroll by rotating the ball.

TIP

Laptops offer a built-in touchpad, which is a flat rectangle in front of the physical keyboard. You can move a finger around the touchpad to move the mouse cursor around your screen. Depending on the touchpad's settings, you may be able to tap on the pad in different spots to simulate clicking or double-clicking. There may also be physical buttons adjacent to the touchpad for clicking and right-clicking.

A **touchscreen** is a display that functions both as output and input. The output is the picture you see on the screen, and the input happens when you drag or tap your finger on items on the screen to do things with them. Windows supports touchscreens, so if your computer has one, Windows can accept commands from it.

If you do own a touchscreen computer or tablet device, you can touch the screen with a finger and then drag your finger across the screen as an alternative to moving the mouse. You can tap or double-tap the screen to simulate clicking or double-clicking. Windows also offers an onscreen keyboard that touchscreen users can optionally use to enter text by tapping the keys.

To "right-click" with a touchscreen, press and hold your finger on the item you want to right-click until a menu appears.

You can also use your finger to **swipe** to the right, left, up, or down. To swipe, touch your finger on a certain spot and then drag your finger across the screen. Swiping is commonly used to move from one item to another (for example, from one photo to the next in the Photos app) or to move up or down on a page.

Windows also offers some touchscreen **gestures** you can make with two fingers at a time, such as pinching. To **pinch**, touch two fingers on the screen apart from each other and then drag the fingers together. Pinching zooms out (shrinks the view). Unpinching is the opposite; it zooms in (enlarges the view). To **unpinch**, start with two fingers together and drag them apart.

Get Familiar with the Start Menu

The Start menu provides access to all the installed apps and many of the settings for your computer. Click the Start button to display the Start menu.

The Start menu looks very different in Windows 11 and Windows 10, so let's have a look at each version separately.

As shown in Figure 2-7, the Start menu in Windows 11 has a Pinned section at the top, containing shortcuts to certain applications. Windows starts out with a default set of pinned apps, but you can pin and unpin items as you wish. Chapter 3 explains how. Unpinning an icon doesn't remove the program; you can still run unpinned programs too.

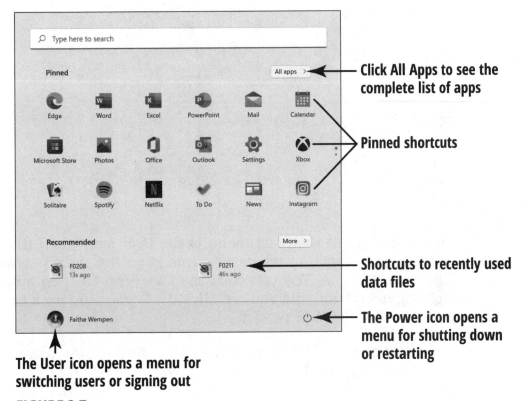

Click All Apps to see the complete list of apps

Pinned shortcuts

Shortcuts to recently used data files

The Power icon opens a menu for shutting down or restarting

The User icon opens a menu for switching users or signing out

FIGURE 2-7

To browse the complete list of available programs, including the unpinned ones, click the All Apps button. Figure 2-8 shows this list.

The lower part of the Start menu is a Recommended section, where shortcuts appear to recently accessed data files you might want to reopen.

FIGURE 2-8

At the bottom of the Start menu is the User icon (the little picture next to your name, or whatever name is on the signed-in account) and a Power icon. You can click the User icon to open a menu for signing in and out, and you can click the Power icon for a menu that lets you shut down or restart.

The Start menu in Windows 10 has two main sections, as you can see in Figure 2-9. On the left is an alphabetical list of all the installed applications. (This is like the list you get in Windows 11 when you click All apps.) You can scroll through this list and click the app you want to run. On the right is a panel where you can pin certain applications for faster access to them. This is like the pinned icons at the top of the Windows 11 Start menu.

The Windows 10 Start menu also has a small column of icons on the far left side. Each of these icons is a shortcut to some special content. When you point to one of these icons, the names of each icon appears.

Alphabetical list of apps   **Pinned shortcuts**

FIGURE 2-9

Here are the icons shown in Figure 2-10, from top to bottom. I explain more about each of these options later.

» **User:** The name of the signed-in user. Click this icon to open a menu where you can switch users, lock the desktop, and sign out. This is equivalent to clicking the User icon at the bottom of the Windows 11 Start menu.

» **Documents:** A shortcut to the signed-in user's Documents folder. There is no direct Windows 11 equivalent for this.

» **Pictures:** A shortcut to the signed-in user's Pictures folder. There is no direct Windows 11 equivalent for this.

» **Settings:** Opens a utility where you can change all sorts of Windows settings. A *lot* more on this later! This is equivalent to clicking the Settings icon (which looks like a gear) on the taskbar.

» **Power:** Opens a menu where you can put Windows to sleep, shut it down, or restart it. This is equivalent to clicking the Power icon at the bottom of the Windows 11 Start menu.

FIGURE 2-10

Sign Out and In

Signing out of your account closes all open files and applications but leaves the computer up-and-running, so that someone else can sign in.

To sign out, follow these steps:

1. **Click the Start button.**

The Start menu opens.

2. **Click the User icon.**

A menu appears.

3. **Click Sign Out.**

You are signed out. A graphic appears with the current date and time on it.

You are prompted to sign in when you start up your computer, and also after you have signed out or chosen to switch users.

To sign in, follow these steps:

1. Press any key to display the sign in prompt.

By default, the prompt appears for the password of the user who most recently signed in.

2. If your account is not the one shown in the center of the screen, click your account name from the list in the lower-left corner of the screen.

The prompt in the center of the screen changes to ask for the chosen user's password or PIN.

3. Enter your password or PIN.

If you enter a password, press Enter afterward. If you enter a PIN, you don't have to press Enter.

Switch Accounts

To allow someone else to sign into your computer, you can either sign out or switch accounts. Signing out closes all open applications and files; when you sign back in again later, you have to re-open everything you want to continue working on. Switching user accounts doesn't sign you out; it just temporarily suspends what you're doing. All the apps and files stay open, and when you switch back, everything is just where you left it.

To switch user accounts, follow these steps:

1. Click the Start button.

The Start menu opens.

2. Click the User icon.

Its name appears as your name, since you are the currently signed-in user.

A menu appears with these options on it: Change account settings, Lock, and Sign out. All of the other user accounts set up on this computer also appear as menu items. In Figure 2-11, for example, there is one other user, that I've generically called Student Name. Figure 2-11 shows Windows 11, but the same menu options appear in Windows 10.

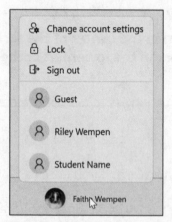

FIGURE 2-11

3. **Click the user name that you want to switch to.**

   Your current session is suspended and a prompt appears for that user to enter their password or PIN.

4. **Type the password or PIN for that account.**

   If you enter a password, press Enter afterward. If you enter a PIN, you don't have to press Enter.

When you're done working with that account and you want to switch back, repeat these steps.

Lock Windows while You're Away

Do you live in a house with a bunch of nosy people who are always trying to see what you're doing on your computer? Windows has you covered. When you get up to go get yourself a sandwich, you can lock your computer from prying snoops without having to sign out.

To lock your computer, follow these steps:

1. Click the Start button.

The Start menu opens.

2. Click the User icon.

A menu appears.

3. Click Lock.

Your Windows session is now obscured from view by an attractive graphic and the current date and time.

4. When you return to your desk, press any key on the keyboard.

A sign-in prompt appears.

5. Re-enter your password or PIN to regain access to your PC.

If you enter a password, press Enter afterward. If you enter a PIN, you don't have to press Enter.

Restart Windows

Sometimes when your computer starts acting strangely — like running more slowly than normal or showing you error messages — you can fix the problem by restarting Windows. Restarting is also known as rebooting. It usually takes less than a minute, and its power to clear up annoying glitches is astounding.

To restart Windows, follow these steps:

1. Click the Start button.

The Start menu opens.

2. **Click the Power icon.**

A menu appears. Figure 2-12 shows the Windows 11 version; the Windows 10 menu is similar.

FIGURE 2-12

3. **Click Restart.**

In a minute or so, a graphic appears with the date and time on it.

4. **Press Enter.**

A sign-in prompt appears, asking for your password or PIN.

5. **Enter your password or PIN to sign back in.**

Place the Computer in Sleep Mode

When you are going to be away from your computer for awhile (with "awhile" being anything from a few hours to a few days), you might want to shut it down completely to save power. However, shutting it down completely also closes all open files and applications. If you would prefer to stay signed in while you're away, but still save electricity, you can put the computer in Sleep mode.

Sleep mode turns off the power to most of the computer's components, but leaves the memory powered up so it can retain what's being stored in it. This dramatically reduces the amount of electricity the computer requires, but still allows you to quickly resume your work.

To place the computer in Sleep mode, follow these steps.

1. **Click the Start button.**

   The Start menu opens.

2. **Click the Power icon.**

   A menu appears.

3. **Click Sleep.**

   The computer appears to be powered off, but is actually just sleeping.

When you are ready to wake the computer up again, try pressing a key on the keyboard. If that doesn't wake it up, press the computer's power button to wake it up.

Shut Down Your Computer

Shutting down the computer stops it from drawing electricity, saving on your electric bill. It also completely shuts down your Windows session, closing all files and applications. You might shut down your computer if you are going to be gone for several days or more, or if you are going to move the computer to a different location.

To shut down the computer, follow these steps:

1. **Click the Start button.**

   The Start menu opens.

2. **Click the Power icon.**

   A menu appears.

3. **Click Shut down.**

   The computer powers off completely.

To restart the computer after shutting it down, you must press the computer's Power button.

Don't simply turn off your computer at the power source unless you have to because of a serious malfunction such as the computer not responding to any commands for more than a minute or so. Windows might not start up properly the next time you turn it on if you don't follow the proper shutdown procedure.

Create Additional User Accounts

During the initial Windows setup, which I walked you through earlier in this chapter, you were prompted to sign in with a Microsoft account. (You might have had to create one, and hopefully you did.) That Microsoft account was set up as an authorized user on your PC.

If it's only you using the computer, then you might not need any more user accounts. But if you are sharing the computer with someone else (or multiple someones), keep reading.

If you're brand-new to Windows as of this chapter, this process of creating new user accounts might be out of your comfort zone at the moment. Feel free to skip this whole section and come back to it later after you've gotten more confident with Windows.

Each user account you set up in Windows must have a Microsoft account. (Well, technically there *is* a way to bypass that requirement and create a local-only account for Windows, but I don't recommend it, because local accounts miss out on a lot of privileges.)

If you have Internet access on your computer, you can create a new Microsoft account by following these steps:

1. **Click the Edge shortcut icon on the taskbar.**

The icon looks like a blue and green swoopy circle. Microsoft Edge opens, which is the default web browser application in Windows.

2. **Click in the Address bar at the top of the window and type the following:** https://signup.live.com/signup.

Be careful to avoid typos.

3. **Press the Enter key on the keyboard.**

   The Create Account screen appears. See Figure 2-13.

   From here, the path diverges depending on whether you want to use an existing email address or get a new email address.

4. **Do one of the following:**

   - To use an existing email address, click in the `someone@example.com` box and type the address. Then, click Next and follow the prompts to finish setting up the Microsoft account.

   - To get a new email address, click Get a New Email Address, and then follow the prompts to set that up.

When you finish up with that process, you'll have a Microsoft account that you can set up as a user on your Windows PC.

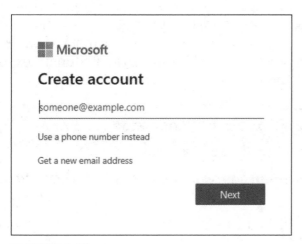

FIGURE 2-13

Use the following procedure to set up new accounts for adults on your PC. These steps assume that the person already has set up their email address as a Microsoft account:

1. **Make sure you are signed in with an administrator account.**

   If you only have one account so far on your PC, it's an administrator account. The next section explains the different types of accounts.

2. **Open the Settings app.**

   One way to do that is to right-click the Start button and choose Settings.

   The Settings window opens.

3. **Click Accounts.**

   Your personal account information appears, including your name and your account type. (It should say Administrator under your email address.)

4. **In the navigation bar on the left side of the Settings window, click Family & Other Users.**

   In Windows 11, it's one of the categories on the right. In Windows 10, it's in the navigation bar on the left.

   You can set up an account as part of your family if you like, but I normally set up other adults as non-family members. The main benefit of setting up another adult as a family member is that you can enable them to control the permission settings for child accounts. I explain how to set up a child account in the next section.

5. **Under the Other Users heading, click Add Account (Windows 11) or Add Someone Else to This PC (Windows 10).**

   The How Will This Person Sign In? prompt appears.

6. **Enter the person's email address associated with their Microsoft account.**

   You do not have to know the person's Microsoft account password.

7. **Click Next.**

   A Good to Go! message appears.

8. **Click Finish.**

   Now this person can sign into your PC with their Microsoft account.

Change an Account's Type

There are two basic types of user accounts in Windows:

» **Administrator**: Has full rights and permissions to make all kinds of changes to Windows, including changes that affect other user accounts.

» **Standard**: Can only make changes to Windows that don't affect other user accounts. For example, they can change the desktop color and the mouse sensitivity, but they can't install or remove software, and they can't view or change other user accounts' personal files.

The first account you create on your PC is automatically set to be an administrator account because you need at least one administrator account in order for Windows to function. So, whatever account you used for the initial setup, that's your administrator account. If you absolutely *have* to change that initial account to a standard account, you'll need to create another administrator account first.

Why the two types? Two main reasons. One is to prevent clueless/careless people from messing things up for others. For example, your visiting great-nephew can't "accidentally" delete all your photos to free up room for him to install his favorite game. The other is to mitigate the damage that computer viruses and other malicious software (malware) can do if your computer gets infected.

Standard user accounts can further be classified as child or adult accounts. You can regulate a child account, restricting what that user can and can't do on your computer.

Did you notice that when you created the new account in the preceding section, you weren't prompted for an account type? That's because by default all new accounts (except the first one) are set to the Standard type. That's for safety's sake.

If you want another user to have administrator privileges, follow these steps to change their account type.

1. **Follow steps 1–4 of the steps in the Create Additional User Accounts section.**

   In the Other Users section you see a list of the other user accounts you've set up.

2. **Click the user account for which you want to change the account type.**

   A Change account type button appears.

3. **Click Change Account Type.**

   The Change Account Type dialog box opens.

4. **Open the Account Type drop-down list and choose the desired type.**

   For example, choose Administrator to bump up the account's privileges, as shown in Figure 2-14.

5. **Click OK.**

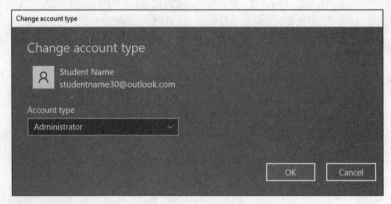

FIGURE 2-14

Manage Family Settings

You can set up accounts to be part of your "family" in Windows, and they don't even need to be blood relatives! Someone you designate as an Organizer family member can manage settings for child accounts in your family. Someone you designate as a Member has fewer privileges in the family's settings.

Follow these steps to create a new account for a child or adult family member:

1. **Open the Settings app.**

   In Windows 11, you can get there by clicking the Settings icon (the gear) on the taskbar. In Windows 10, click the Start button and then click Settings.

2. **Click Accounts.**

   Your personal account information appears, including your name and your account type. (It should say Administrator under your email address.)

3. **Click Family & Other Users.**

   In Windows 11, it's one of the categories on the right. In Windows 10, it's in the navigation bar on the left.

4. **Under the Your Family heading, click Add Account.**

   An Add Someone prompt appears.

5. **In the Enter Their Email Address box, type the person's email address that they use for their Microsoft account and click Next.**

   If they don't have a Microsoft account, you can click the Create One for a Child link and follow the prompts to set one up.

6. **When prompted about what role they should have, click either Organizer or Member.**

   Organizer would be appropriate for another adult in the family that you trust; Member would be appropriate for a child, or for an adult you might not trust completely.

7. **Click Invite.**

   The person's email address appears under the Your Family heading in the Settings window. To complete the setup process, that person needs to respond to an invitation sent to their email inbox.

After setting up one or more family members (as in the preceding section), you can return to the Settings window and under Your Family, click Manage Family Settings online to open the Microsoft Family Safety web page. From here, you can sign in with your Microsoft account (click Sign In Now) and then review and change the settings for your family. For example, in Figure 2-15, I'm looking at the options for family member Margaret; I can check her screen time, set content filters, and more.

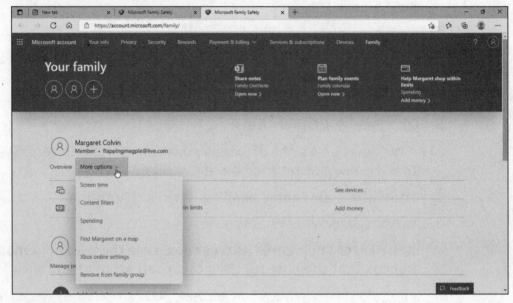

FIGURE 2-15

IN THIS CHAPTER

» **Determining whether you need a printer**

» **Choosing the right printer**

» **Setting up a new printer**

» **Connecting a printer to your PC**

» **Setting a default printer**

» **Setting printer preferences**

» **Managing a print queue**

» **Removing a printer**

Chapter **3**

Buying and Setting Up a Printer

A computer is a great storehouse for data, images, and other digital information, but sometimes you might need to print hard copies (a fancy term for paper printouts) of electronic documents and images. For example, if you need to bring some notes to a meeting, it's a lot easier to bring a printed sheet of paper than to lug your computer along with you.

A **printer** enables you to create hard copies of your files on paper, transparencies, or whatever materials your printer can accommodate. This chapter provides advice for deciding which printer to buy, setting it up, and adjusting its settings.

Do You Need a Printer?

Let's kick off the topic of printers by considering whether it's even relevant to your situation. Not everyone needs a printer, just like not everyone needs an automobile or a smartphone. (And if you're recoiling in horror at the thought of not having a car and a smartphone, forget I even said that.)

A decade or so ago, I would have said that almost everyone with a computer needs a printer. There were so many situations that required you to print hard copies back then, from concert tickets purchased online to your tax returns filed using tax software. Today, though, the world is changing. Rather than assuming that everyone has a printer, it's more common to assume that everyone has a smartphone. If you buy concert tickets online these days, you will probably get an e-ticket that you display on your phone at the venue gate. And if you file your taxes using tax software, you can create a PDF version of your taxes to store electronically for your records. No filing cabinet required!

So, do you need a printer? Maybe, maybe not. You might want to wait for a few months after buying your first computer to see if any situations come up where you wish you had one. You can use a printer at a FedEx or UPS store for a nominal charge, and you probably have a friend or two who has a printer you can use. If you do decide that having your own printer makes sense, check out the next section. Otherwise, we're done here, and you can move on to Chapter 4.

Choose the Right Printer

There are two choices of printer technology today for the home user: inkjet and laser. An **inkjet printer** uses liquid ink squirted out of tiny jets onto the paper. A **laser printer** uses powdered toner like a copier does.

First, let's talk cost. An inkjet printer is inexpensive initially (usually under $100), but the ink cartridges are rather pricey (about $35 for a cartridge that you'll maybe get 200 pages out of, and each printer has four different colors of ink to contend with).

A laser printer is more costly initially $150 and up for a black-and-white model, $400 and up for color), and the toner cartridges are also expensive ($60 or more per color). However, toner cartridges last a lot longer; you can expect to get 2,500 pages or more before you have to change the toner cartridge. If you do a lot of printing, the price difference per page can really add up.

In terms of quality, a laser printer typically is better. Inkjet ink can smear if you touch it immediately after it comes out of the printer, and the jets on the ink cartridge tend to get clogged with dried ink if you don't use the printer frequently, causing you to have to run a cleaning cycle that uses up some of the ink.

REMEMBER

There's one area in which inkjets are superior in quality, though, and that's photo printing. Some inkjet printers are designed to excel at photo printing on special glossy paper, and laser printers can't compete with that because they can't print on that special paper. (The special paper is expensive, though, like everything else that an inkjet printer consumes. Expect to pay about 50 cents per letter-sized sheet.)

Laser printers are also handier if you need to print many pages at a time, because they typically have larger paper trays, so you have to load paper less often. Some of them even have multiple paper trays.

If space is an issue, make sure you are aware of a printer's footprint (size) before you buy it. In general, inkjet printers tend to be smaller than lasers, but it varies by model. Some laser printers are quite big and bulky, so check the specs and measure the available space to make sure it's a fit.

TIP

Some printers have multiple functions. In addition to printing, they also copy, scan, and some of them even fax (if you hook them up to a phone line). Such printers are often referred to as **multifunction devices** (MFDs). They cost a bit more, but if you were going to buy all those extra devices anyway, it's a net savings of both money and space to have them all in one device.

You should also consider printer sharing if you have more than one computer in your home. One way to share a printer is to hook it up directly to one of the computers and then have that computer share its printer with other computers on the same home network. (Obviously you need a home network set up for that to work.) This method is not optimal because it works only when both computers are up-and-running. An alternative is to look for a printer that has a built-in network interface. You can then hook up the printer directly to your network's router and all the computers on the network can use it independently of one another. A printer's network interface can be either wired (requiring a cable) or wireless.

Unpack and Install a New Printer

As you unpack the printer, carefully remove any tape on the printer; it's common for a printer to come with a half-dozen or more pieces of tape holding various compartments closed for shipping. Save all the packing materials, especially the Styrofoam parts that protect the printer from impact. If you ever need to move the printer, you'll be glad you can repack it safely.

Find the setup directions that came with the printer and follow them to the letter. These directions explain how to install the ink or toner cartridges, how to print a test page, and how to connect the printer to a computer or to your network. Nearly all printers today connect via either a USB cable (directly to a computer) or via a network interface (either a network cable to a router, or a wireless network connection to a router).

Set Up a Printer to Work with Windows

When you connect a new printer directly to your Windows PC with a USB cable, Windows should automatically detect it and try to install a driver for it. A **driver** is a small computer program that translates between the operating system and the printer's hardware. Windows comes with a small library of drivers for the most popular printers, and if you have Internet access, Windows can also connect to a larger library of drivers online to find the right one. This process is called **Plug and Play (PnP)**, and it's pretty much automatic; you don't usually have to do anything.

The drivers that Windows sets up via PnP give the printer basic functionality. In other words, it sets up the printer to be able to print from any Windows app. Most printers also have extended features and capabilities that you can use if you have special software provided by the printer manufacturer, such as monitoring the ink/toner levels and popping up messages to let you know when the paper is jammed. MFD printers also have software that controls its scanning and copying capabilities. To have all that extra stuff, though, you have to install the manufacturer's own software.

In earlier days a printer would come with a CD or DVD containing the manufacturer's software, including the drivers. Nowadays, though, a lot of computers don't have a CD/DVD drive, so instead of providing a disc, many printer manufacturers instead provide a flyer in the box with the printer that lists an Internet address where you can download the printer software. Follow the instructions.

If Windows and the printer are both operating correctly, and directly connected via a USB cable, they should recognize each other automatically. If they don't, or if you need to set up a printer that's connected to your network router rather than to any individual PC, you can ask Windows to find the new printer by following these steps:

1. **Open the Settings app.**

   One way is to right-click the Start button and choose Settings.

2. **In Windows 11, click Bluetooth & Devices ⇨ Printers & Scanners ⇨ Add Device.**

   Or

   In Windows 10, click Devices ⇨ Printers & Scanners ⇨ Add a Printer or Scanner.

   Windows searches for any available devices.

3. **If the desired printer appears on the list of search results, click it and then click Add Device.**

See Figure 3-1.

Windows should set up the new printer automatically from this point, downloading a driver for it if needed. If any other prompts appear, respond to them to complete the setup.

Search results

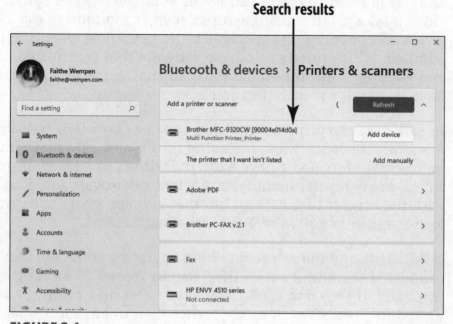

FIGURE 3-1

If the preceding procedure doesn't work, there's probably something wrong. Maybe the printer is not powered on, or not connected to your network or PC, or the ink or toner cartridges aren't properly installed. (Maybe you left a piece of tape on one of them by accident.) Check the printer's setup instructions to see where you might need to perform an extra step or two.

There is a manual process you can go through to set up a printer that refuses to be automatically seen in Windows, such as an older printer or one that uses a Bluetooth interface. Click Add Manually (Windows 11) or The Printer That I Want Isn't Listed (Windows 10), and then work through the Add Printer wizard's steps, as shown in Figure 3-2.

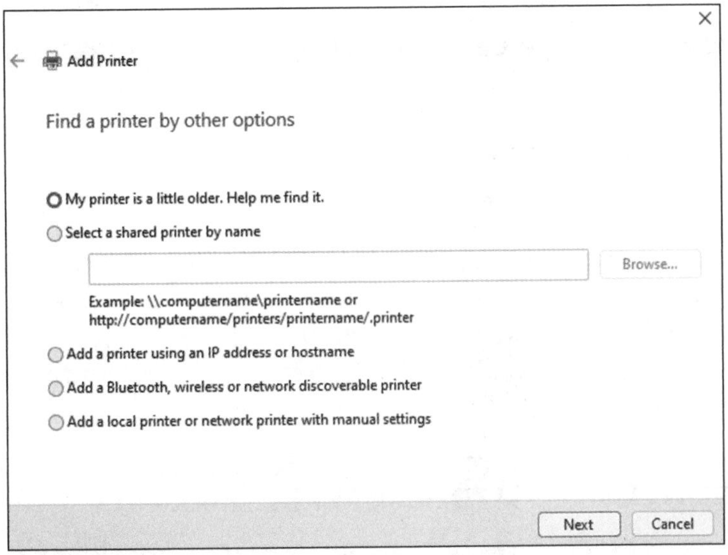

FIGURE 3-2

Set a Default Printer

If you have only one printer connected to your PC (or available via your network), then it's automatically the default printer. If there's more than one, though, you can choose which one is the default. The default printer is the one that gets used when you don't specify a certain printer for a particular print job.

Windows has a feature called Let Windows Manage My Default Printer. When it's enabled, you won't be able to manually choose which printer is the default; Windows will make that call based on the one you used most recently at your current location. You have to turn off that feature in order to make a specific selection.

The following steps show how to manually choose your default printer.

1. **Open the Settings app.**

   One way is to right-click the Start button and choose Settings.

2. **In Windows 11, click Bluetooth & Devices ⇨ Printers & Scanners.**

   Or

   In Windows 10, click Devices ⇨ Printers & Scanners.

   A list of the installed printers appears.

WARNING

   Even if you only have one physical printer, there may be multiple printing devices listed under Printers & Scanners. Some software appears in Windows to be a printer, such as Microsoft XPS Document Writer and Adobe PDF. These non-printer "printers" can also be set as your default printer.

3. **Scroll down to the bottom of the list of printers and locate the Let Windows Manage My Default Printer check box.**

4. **If the feature is set to On (Windows 11) or if its check box is marked (Windows 10), set it to Off or clear the check box.**

   Otherwise, you won't be able to choose which printer should be the default. See Figure 3-3.

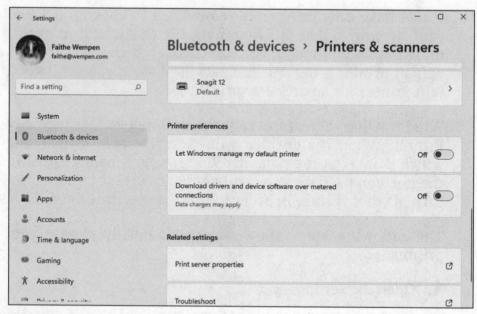

FIGURE 3-3

5. **Scroll back up the list if needed, click the printer that should be the default.**

6. **If you're using Windows 10, click Manage.**

   Skip this step if you're using Windows 11.

7. **Click the Set as Default button.**

   If there is no Set as Default button, you did not clear the check box in step 3.

8. **Click the Close (X) button in the upper-right corner of the Settings app window.**

Set Printer Preferences

Your printer might have capabilities such as choosing whether to print in color vs. black and white, or choosing whether to print in draft quality (which uses less ink) or high quality (which produces a darker, crisper image). It also might let you choose a default paper tray (if it has more than one) and an orientation (portrait or landscape).

To modify these settings for all documents you print, follow these steps.

1. **Open the Settings app.**

   One way is to right-click the Start button and choose Settings.

2. **In the Windows 11 Settings app, click Bluetooth & Devices ⇨ Printers & Scanners.**

   Or

   In the Windows 10 Settings app, click Devices ⇨ Printers & Scanners.

   A list of the installed printers appears.

3. **Click the printer for which you want to set preferences.**

4. **If you're using Windows 10, click Manage.**

   Skip this step if you're using Windows 11.

5. **Click Printing Preferences.**

The Printing Preferences dialog box opens for that printer. The controls in that dialog box depend on the printer driver; the available settings are different for different types and brands of printers. Figure 3-4 shows one example.

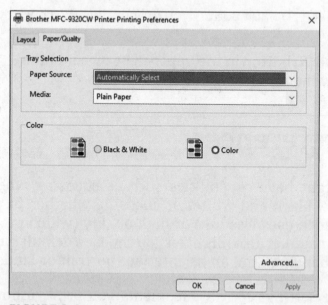

FIGURE 3-4

6. **Click any of the tabs to display various settings, such as Paper/Quality.**

Note that different printers might display different choices and different tabs in this dialog box, but common settings include:

- **Color/Black & White:** If you have a color printer, you have the option of printing in color. The Black & White option uses only black ink. Some printer models call that mode **grayscale**. When printing a draft of a color document, you can save colored ink by printing in black and white or grayscale.

- **Quality:** Some printers allow you to print in fast or draft quality (these settings may have different names depending on your printer's manufacturer) to save time and/or ink, or print in a higher or best quality for your finished documents. Some printers also offer a dpi (dots-per-inch) setting for quality — the higher the dpi setting, the better the quality.

- **Paper Source:** If you have a printer with more than one paper tray, you can select which tray to use for printing. For example, you might have 8½-x-11-inch paper (letter sized) in one tray and 8½-x-14-inch (legal sized) in another.

- **Paper Size:** Choose the size of paper or envelope you're printing to. In many cases, this option displays a preview that shows you which way to insert the paper. A preview can be especially handy if you're printing to envelopes and need help figuring out how to insert them in your printer.

7. **Click the OK button to close the dialog box and save settings.**

TECHNICAL STUFF

The settings you just looked at are for the printer's internal formatting and handling of print jobs. There is also another set of printer settings that have to do with ports, sharing, security, and other more technical issues that home-based users seldom need to be concerned with. To work with those settings, choose Printer Properties rather than Printing Preferences in Step 3.

8. **When you are finished with the printer settings, click the Close (X) button in the upper-right corner of the Settings app window.**

The Settings app window closes.

TIP

Whatever settings you make using the preceding steps become your default settings for all printing you do, across all applications. However, when you're printing a document from within an application, the Print dialog box that appears there gives you the opportunity to change the printer settings for printing that document only.

Manage a Print Queue

Each printer has its own separate queue in Windows. A **queue** is like a line that jobs wait in to be processed. (You'd know that if you were British.)

When you print something in a Windows app, the print job gets **spooled** (a fancy word for *sent*) to the printer's queue, where it waits its turn to be printed, along with print jobs from any other applications that happen to be printing at that same time. Normally, the queue fills up and then empties behind the scenes, and you don't have to mess with it.

When there are items in the print queue, a printer icon appears in the notification area (the bottom-right corner of the Windows desktop screen). The icon disappears when the last print job is spooled to the printer. That's useful to know because if something goes wrong — that is, if your print job doesn't print — you can look for that icon to help you troubleshoot the problem.

Here's how to troubleshoot a print problem by looking at its queue:

1. **In the notification area, find and double-click the printer icon. (See Figure 3-5.)**

The printer's queue window opens.

Printer icon

FIGURE 3-5

2. **Look for a reason why the print job is not completing.**

For example, in Figure 3-6 the queue shows Paused. That's probably why it's not printing. To fix that, click Printer to open the Printer menu and click Pause Printing to toggle the pause off.

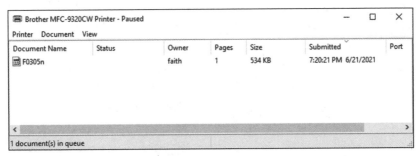

FIGURE 3-6

You can also control the print jobs from the queue window. For example, select a waiting print job, then click Document to open the Document menu and choose Pause to pause one particular job. Choose Resume to resume a paused job.

To cancel a print job before it is sent to the printer, select it and press the Delete key on the keyboard, or open the Document menu and choose Cancel.

Remove a Printer

Over time, you might upgrade to a new printer and chuck the old one. When you do, you might want to remove the older printer driver from your computer. That takes it off the list of printers that you see in applications.

To remove a printer's driver from Windows, follow these steps:

1. **Click the Start menu and then click Settings.**

The Settings app opens.

2. **Click Devices ⇨ Printers & Scanners.**

The Printers & Scanners list appears.

3. **Click the printer and then click the Remove Device button.**

A confirmation prompt appears. See Figure 3-7.

4. **Click Yes.**

The printer is removed from the printers list.

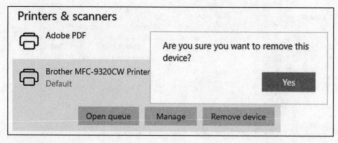

FIGURE 3-7

TIP

If you remove a printer, it's removed from the list of installed printers, and its driver is uninstalled. If it was the default printer, Windows assigns the default status to another printer (if there is one). You can no longer print to the removed printer unless you install it again.

WARNING

Even after you follow the process just described, there might still be some remnants of the printer's software installed on your PC. Chapter 4 explains how to uninstall unwanted software; put what you learn there into practice by removing any apps associated with the no-longer-installed printer.

2

Getting Up to Speed with Windows

IN THIS PART . . .

Working with Windows applications

Exploring some of the apps that come with Windows

Managing your personal files

Customizing Windows

IN THIS CHAPTER

» Learning the names of things in Windows

» Starting an app

» Exiting an app

» Finding your way around in a desktop app

» Finding your way around in a Microsoft Store app

» Working with a window

» Switching among running apps

» Moving and copying data between apps

» Installing new apps

» Removing apps

Chapter **4**

Working with Apps in Windows

Most of the useful and fun things a computer can do require you to interact with the operating system, Microsoft Windows. Through Windows, you can run business applications and games, organize your personal files, and more.

This chapter officially kicks off your Windows orientation. Here you explore the Windows desktop interface and learn some basic skills that you use in every session. I start by teaching you the names of the parts of the Windows interface, so when you read the directions for specific tasks later, you'll know what to do. You learn how to start and exit applications, how to navigate an app's controls, and how to move and resize their windows. You also learn how to switch between the running applications and how to install and remove apps from Windows.

Learn the Names of Things

As you work through this book, there will be lots of steps for you to follow to complete specific tasks. You've already seen examples of that in the first three chapters, and there's much more to come! To follow along with these steps, you need to understand some basic Windows vocabulary.

Figure 4-1 shows the Windows 11 desktop and points out some of the key features, which are defined in Table 4-1. Figure 4-2 points out the equivalent of these same features in Windows 10. If you don't know what some of these things are, don't stress — you'll learn about them later.

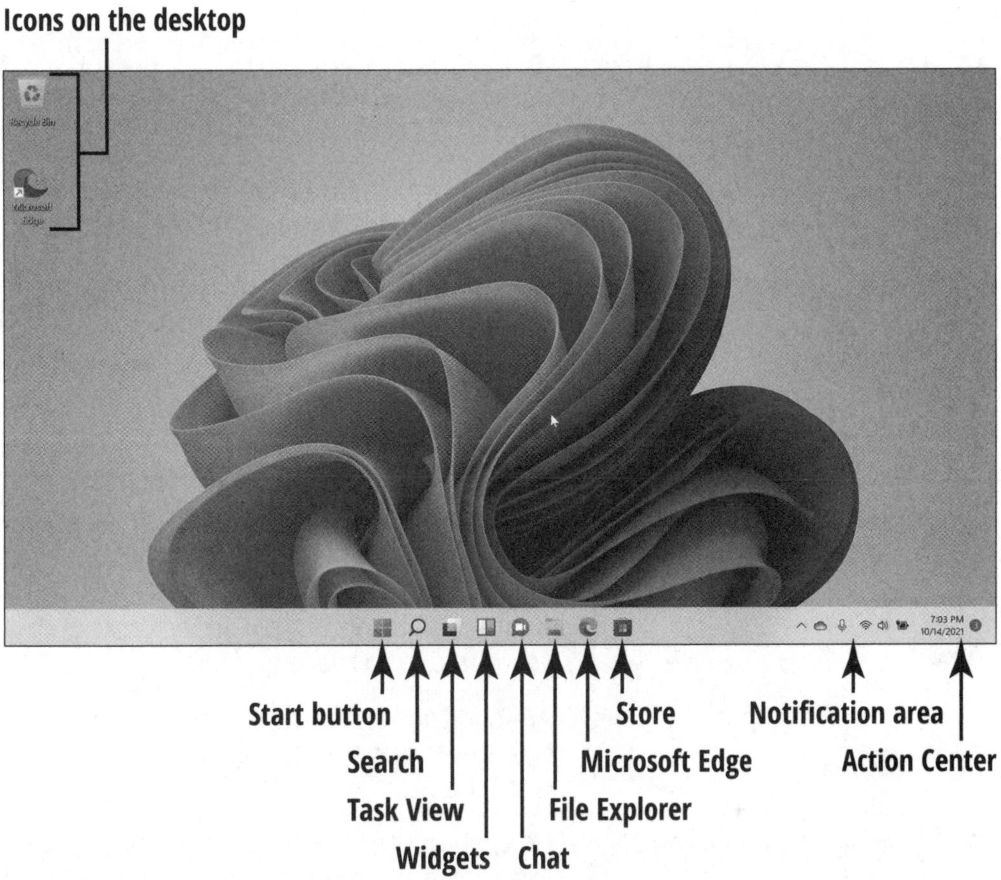
Icons on the desktop
Recycle Bin
Microsoft Edge
7:03 PM
10/14/2021
Start button
Search
Task View
Widgets
Chat
File Explorer
Microsoft Edge
Store
Notification area
Action Center

FIGURE 4-1

TABLE 4-1 **Some Key Windows Features**

Icon	Name	Description
	Icon	A picture that represents a shortcut to a file, application, or Windows feature.
	Start button	The button you click to open the Start menu.
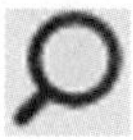	Search	A text box (Windows 10) or a button that opens a text box (Windows 11) where you can type words and phrases you want to search for.

(continued)

TABLE 4-1 *(continued)*

	Cortana (Windows 10 only)	An icon that opens Cortana, a personal assistant application that can perform useful services for you.
	Task View	A Windows feature that helps you manage your running applications.
	Widgets (Windows 11 only)	An icon that opens a gallery of widgets (small desktop apps) you can choose to appear on your desktop.
	Chat (Windows 11 only)	An icon that opens a chat program you can use to communicate with people via typed instant messages.
	Microsoft Edge	An icon that opens the Microsoft Edge web browser.
	File Explorer	An icon that opens the File Explorer utility, which you can use to manage files.
	Settings (Windows 11 only)	An icon that opens the Settings app, where you can change Windows settings. In Windows 10, this icon is on the Start menu.
	Notification area	An area of the taskbar that contains the icons for apps and utilities that are running in the background. It also shows the current date and time.
	Action Center	An icon that opens the Action Center panel, where you can view and manage system notifications. A number, if shown, represents the number of unread notifications.

Icons on the desktop

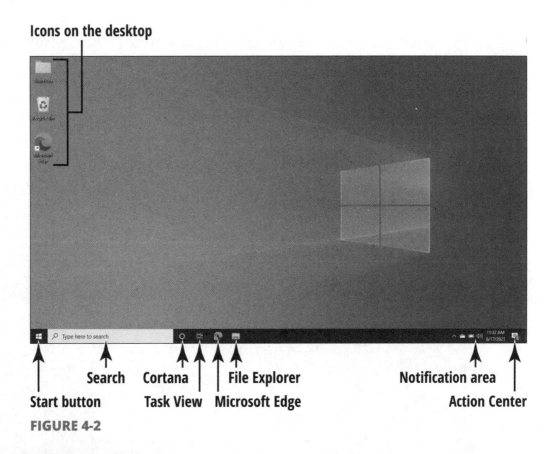

Search **Cortana** **File Explorer** **Notification area**

Start button **Task View** **Microsoft Edge** **Action Center**

FIGURE 4-2

Start an App

There are several different methods you can use for starting an application, which is nice because you can pick the method that you prefer, or that works best in a particular situation. You can

» **Open the Start menu and click the app's name in the alphabetical list.**

In Windows 11, click All Apps to see the alphabetical list; in Windows 10, it's right there when you open the Start menu.

You will probably need to scroll down in the alphabetical list to find the entry you want. To scroll down, position the mouse pointer to the right of the alphabetical list, so a **scroll bar** appears. It's a thin vertical bar with a smaller section at the top. That smaller section at the top is the **scroll box**. See Figure 4-3. Drag the scroll box downward to scroll down the list. Alternatively, click the tiny down-pointing arrow at the bottom of the scroll bar (the **scroll arrow**) to scroll down.

Scroll box   Scroll bar

Scroll arrow

FIGURE 4-3

» **Open the Start menu and start typing the app's name, and then click that name in the search results to open it.**

For example, in Figure 4-4, I typed *Paint*. Windows 11 is shown, but the process is the same in Windows 10; it just looks a little different.

>> **Click the Search button to open a Search pane (Windows 11) or click in the Search box on the taskbar (Windows 10) and start typing the app's name. When the app's name appears among the search results, click it.**

This is basically the same as opening the Start menu and starting to type the app name, because as soon as you do that, the Search box becomes active, like in Figure 4-4.

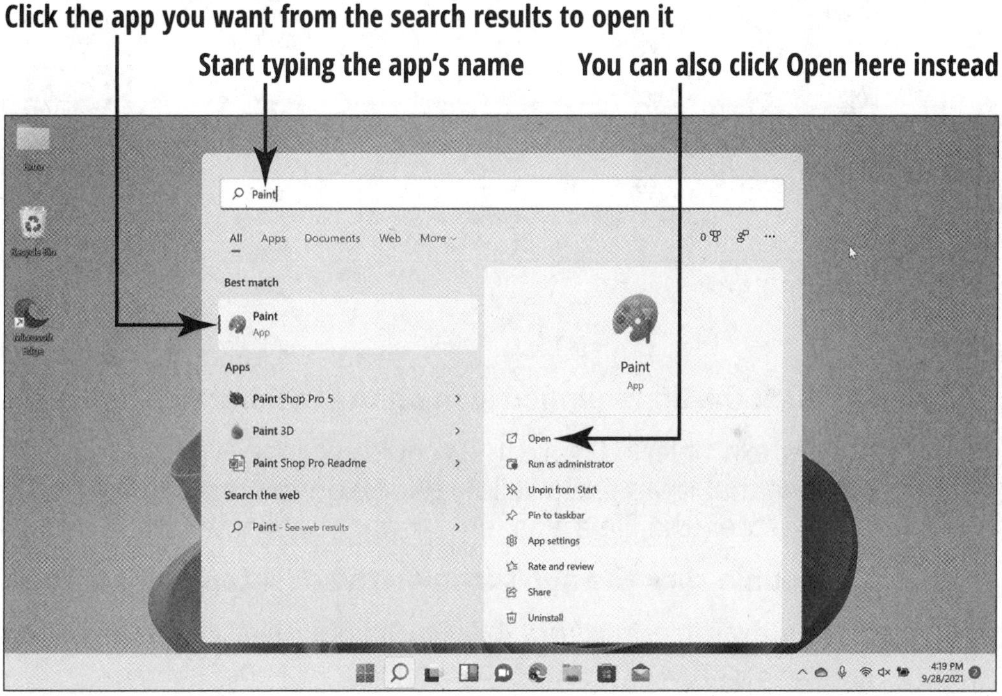

Click the app you want from the search results to open it

Start typing the app's name   You can also click Open here instead

FIGURE 4-4

>> **Click the app's pinned shortcut icon (Windows 11) or tile (Windows 10) on the Start menu, if it's there.**

For example, in Figure 4-5 I could click the Word tile to open that app. In Windows 11, the shortcut icons are all the same size and color; in Windows 10, they may be different colors and sizes, but they work the same way.

Microsoft Edge icon

FIGURE 4-5

» **Click the app's pinned icon on the taskbar (if it's there).**

For example, in Figure 4-5, I could click the File Explorer icon (the icon that looks like a yellow folder) on the taskbar to open File Explorer (the file management app that comes with Windows).

» **Double-click the app's icon on the desktop (if it's there).**

For example, in Figure 4-5, I could double-click the Microsoft Edge icon on the desktop to open that app.

» **Locate the app in the File Explorer utility (if you happen to know where it's stored) and double-click it there.**

That method is the long and hard way around, but sometimes it's the only way (for example, if you've downloaded an app that didn't place an icon for itself on the Start menu's list when you installed it.

Different people have different opinions as to which of these is the best. Try out each one of them and see what you think!

TIP

For all of those "if it's there" options on the preceding list, in Chapter 7 you learn how to make sure the apps you want to run are pinned or placed handily where you want them.

Exit an App

When you open an app, it appears in a well-defined rectangular area called a **window**. A button appears on the taskbar for each open app. You can tell that a button on the taskbar represents an open window or application because it has a horizontal line under it. See Figure 4-6.

Closing an app means closing its window. Here are some ways to do that:

» Right-click the app's taskbar button to display a menu, as shown in Figure 4-6, and then click Close all windows (or Close window).

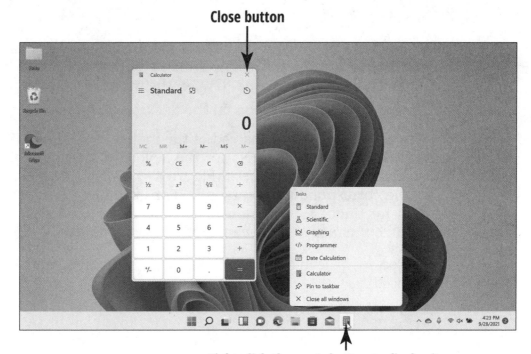

Close button

Right-click the app's button to display its menu

FIGURE 4-6

- » Click the Close (X) button in the upper-right corner of the app's window.

- » If the app has a menu system that includes an Exit command, select it. (You learn about menu systems later in this chapter.)

Find Your Way Around in a Desktop App

Different apps have different ways you can control and communicate with them. There are two basic types of apps that Windows can run: desktop apps and Microsoft Store apps. We look at desktop apps in this section, and Microsoft Store apps in the next section.

Desktop apps are traditional apps designed primarily for use on desktop PCs. They are the kind of apps that Windows has supported for decades. Because they have been around for so long, different interface styles have developed for them. That means I can't explain just one method of navigating a desktop app here; I have to show you several versions.

Older desktop apps have a traditional menu system that consists of a **menu bar** across the top of the window with text names for each menu. You click a menu name to open the menu, as shown in Note-pad in Figure 4-7.

TIP

You can use the keyboard instead of the mouse to navigate in an application if you prefer it. Notice the keyboard shortcuts listed next to each command name in Figure 4-7. Press those keys to select the corresponding commands. For example, Ctrl+Z is the shortcut for Undo. Most Windows apps have keyboard shortcuts for most commands, for convenience and accessibility.

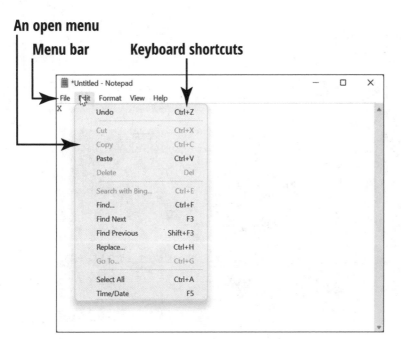

An open menu

Menu bar   **Keyboard shortcuts**

FIGURE 4-7

Some apps have one or more **toolbars**, which are rows of icons (little pictures) across the top or side of the application window. Each button is a shortcut to a particular command. Microsoft Edge is an example of this type of app interface, with a toolbar across the top. Its toolbar contains some buttons for different commands, and also an Address bar, where you type the address of the website you want to visit. Notice the More button (. . .) in the upper-right corner. In many toolbar-based apps, the More button opens a menu, as shown in Figure 4-8.

TIP

Wondering how to tell what the buttons do on a toolbar, and what the keyboard shortcut equivalents are? Hover the mouse pointer over an icon to see a **ScreenTip**, a pop-up message that tells you both the button's name and the keyboard shortcut for it (if there is one).

More button

FIGURE 4-8

Microsoft has been gradually moving many of its apps to a Ribbon interface. A **Ribbon** is basically a multi-tabbed toolbar; you click different tabs above the Ribbon to switch to different pages of commands. WordPad has a Ribbon interface, for example, as shown in Figure 4-9. The words File, Home, and View across the top are Ribbon tabs. The Home tab is the active one.

TIP

Ribbons have keyboard shortcuts, too. In a Ribbon interface, you can press the Alt key to see ScreenTips that tell you what key you can press to activate each tab. When you press that key (for example, Alt+H for the Home tab), you see more ScreenTips that show what key to press to activate each command. For example, Figure 4-10 shows all the ScreenTips for the commands on the Home tab.

Ribbon

Ribbon tabs

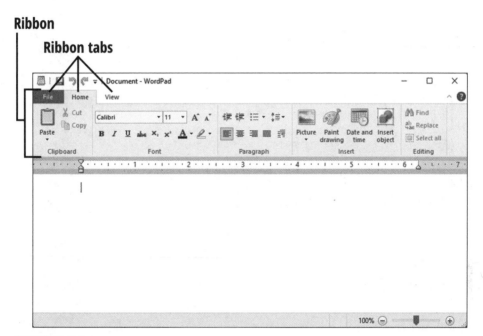

FIGURE 4-9

FIGURE 4-10

Most apps that use the Ribbon style of interface have a **File tab** at the far left; it's often a different color from the other tabs to help distinguish it. Clicking the File tab opens some sort of menu. Depending on the app it might be a two-column menu similar to the one shown in Figure 4-11, or it might be a full-window screen called Backstage view with special commands for managing the app's content, like in most of the Microsoft Office suite applications.

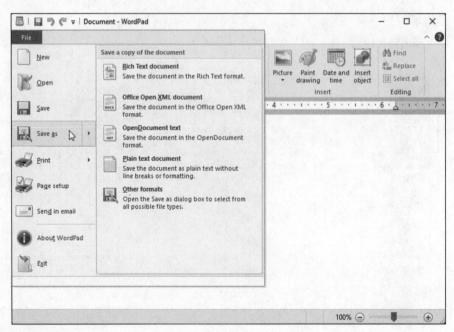

FIGURE 4-11

Find Your Way Around in a Microsoft Store App

Microsoft Store apps are generally simpler and less feature-rich than desktop apps, because they are designed to run well not only on desktop and laptop PCs, but also on mobile devices such as tablets. Their interfaces are optimized for touchscreen use for that reason, using large, simple icons that are widely spaced apart to allow for large fingers to precisely tap one icon or another.

The Weather app in Windows is an example of a Microsoft Store app. In Figure 4-12, notice how the Weather app has a toolbar along the left edge.

The button at the top of the icon bar in Figure 4-12 with the three horizontal lines is a Menu button, but it is often called a "hamburger" button because the three horizontal lines look like a hamburger. This button opens a menu bar that shows the full names of all the icons and also has some commands at the bottom.

Toolbar

Hamburger button

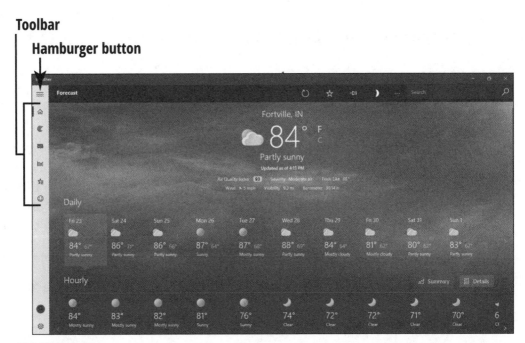

FIGURE 4-12

There's a lot of variation among the various Microsoft Store apps; buttons can appear around any of the four sides of the application window, and there might or might not be a hamburger button.

Work with a Window

The name *Microsoft Windows* comes from the fact that each app runs in its own rectangular area called a **window**. It's important to know how to work with these windows so you can organize the Windows desktop for maximum productivity (or fun) as you are working with applications.

Whenever you start an app, you open a window for it to run inside of; whenever you exit an app, you close its window. (Conversely, whenever you close a window, you exit whatever app was running in it.) The easiest way to close a window is to click its Close (X) button in its upper-right corner.

An open window can have one of three states:

- **» Maximized:** It fills the entire screen.
- **» Restored:** It does not fill the entire screen, but it is visible. It can be resized and moved.
- **» Minimized:** The window is hidden from view, but is still open.

Figure 4-13 shows an application window maximized, Figure 4-14 shows it restored, and Figure 4-15 shows it minimized. Notice that in all three of these figures, you see a button for the app on the taskbar; that doesn't change.

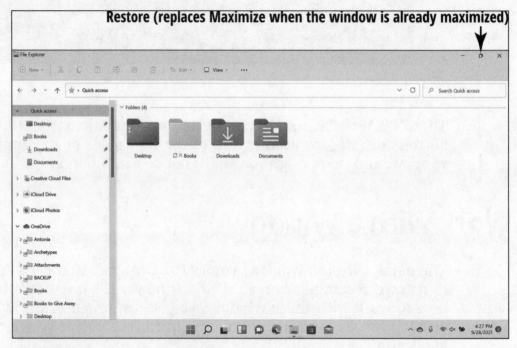

FIGURE 4-13

You can control a window's state using the buttons in its upper-right corner. Table 4-2 shows these buttons and explains what each one does. Depending on the app, the button style may be somewhat different.

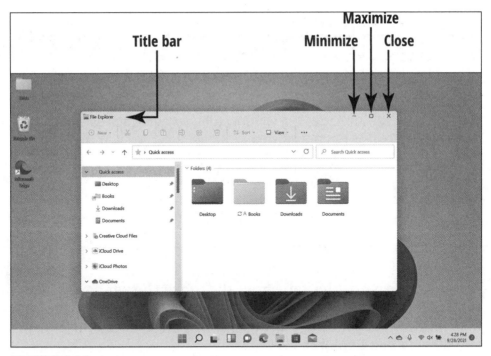

FIGURE 4-14

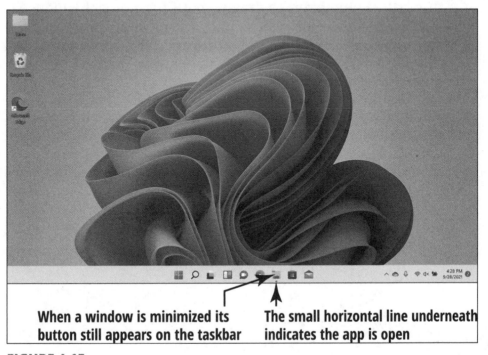

When a window is minimized its button still appears on the taskbar

The small horizontal line underneath indicates the app is open

FIGURE 4-15

TABLE 4-2   **Window Control Buttons**

Button	Name	Purpose
—	Minimize	Changes the window state to *minimized.*
▢	Maximize	Changes the window to *maximized*; appears only if the window is not maximized.
⧉	Restore	Changes the window state to *restored*; appears only if the window is maximized.
✕	Close	Closes the window, exiting whatever application was running in it.

The **title bar** is the bar across the top of the window where the application's name appears, or sometimes the name of the active file or location, depending on the app. In Figure 4-14, the title of the window is File Explorer.

Restored windows can be moved around. To move a window, drag its title bar. (Remember, to drag something, point the mouse pointer at it, hold down the left mouse button, and then move the mouse while continuing to hold down the button.) You might want to move multiple restored windows around to arrange them so you can see both at once, for example.

Restored windows can also be resized. To resize a window, point the mouse pointer at the border of the window (any side except the top), so the mouse pointer turns into a two-headed arrow, as in Figure 4-16. Then drag inward to make the window smaller or outward to make it larger.

Mouse pointer

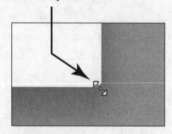

FIGURE 4-16

Switch among Running Apps

Most people run multiple applications at once. For example, you might have your email application open all day long, or a web browser displaying the top news headlines, while you write a story in a word processing program or listen to music using a music application.

Windows provides several different ways for you to switch between applications.

» **The taskbar:** Perhaps the easiest is to just click the button in the taskbar for the app you want to work with. It pops into the foreground, becoming active. The **active window** is the one that is at the top of the stack — the one you are currently working with.

» **Alt+Tab:** Another way to switch applications is to hold down the Alt key and press and release the Tab key. This opens a panel showing thumbnail images (small images) of each open window. (See Figure 4-17.) While continuing to hold down the Alt key, press and release the Tab key again. Each time you press Tab, a colored outline appears around a different thumbnail image. When the app you want has the outline, release the Alt key and that app becomes active.

» **Task view:** The Task View icon is the icon on the taskbar that looks like a black square and a translucent white square (Windows 11), or a filmstrip (Windows 10). A keyboard shortcut for it is Windows key + Tab. When you activate Task view using either method, thumbnail images of each of your open apps appears (see Figure 4-18), and you can click the one you want to activate. This is a lot like Alt+Tab, but you don't have to repeatedly press a key to cycle through the open apps; you can go directly to the one you want.

FIGURE 4-17

FIGURE 4-18

MULTIPLE DESKTOPS

Perhaps you're wondering about the New Desktop option in Figure 4-18. Task View has many features that are more advanced than we have time to get into in this book, and having multiple desktops is one of them. This feature was created to help out people who have so many different apps open at once that their desktop gets crowded and difficult to manage. By creating a new desktop, you can start fresh on a new desktop without having to close any of the apps that are already open. To switch among the different desktops, click the Task View icon or press Windows key + Tab and then click the thumbnail image for the desired desktop at the top of Task View. Be careful when working with multiple desktops, though, because it's easy to forget that you've got multiple desktops going at once and re-open an application in your new desktop that is actually already open in the original desktop.

Move and Copy Data between Apps

Most apps are designed to create and manage some sort of data. For example, word processors, spreadsheets, databases, and graphic arts applications all help you work with data.

You can move and copy data between most applications — even when the data in one application isn't in the same format. For example, you can copy some cells from a spreadsheet into a presentation graphics program, or a photo from a camera app into a word processing application. This is made possible by a Windows feature called the **Clipboard.** The Clipboard is like a holding tank for data. You place some data from one application on the Clipboard, and then switch to another application and paste the data from the Clipboard into that application.

The Clipboard is a built-in part of Windows; it is not an app. You do not have to start or stop it; it's active all the time in the background, just waiting for you to use it.

There are three operations you can perform on selected data that involve the Clipboard:

» **Cut:** Removes the selected data from its original location and places it on the Clipboard.

» **Copy:** Leaves the selected data in its original location and places a copy of it on the Clipboard.

» **Paste:** Inserts a copy of the item on the Clipboard into the active application, at the specified location.

Before you can cut or copy data to the Clipboard, you must **select** the data. Different applications have different ways of selecting data, but in general, you either click it (which works for most graphic objects) or drag across it (which works well for spreadsheet cells and blocks of text). After you make your selection, you issue either the Cut or Copy command, whichever is appropriate to your situation. Then make the destination location active and use the Paste command.

You can place any amount of text or graphics (or a combination of things) on the Clipboard; the only requirement is that you select it all and then issue the Cut or Copy command once for the entire selection. Anything on the Clipboard *stays* on it until you cut or paste something else. You can paste it multiple times, into multiple locations. When you sign out of Windows or shut down your PC, whatever was on the Clipboard is erased.

The Clipboard can hold only one item at a time. That item can be large or small; it could be a single word or a 100-page manuscript or even 100 different files. It doesn't matter — if you selected all the things and then put them on the Clipboard in one operation, then the group of things is considered a single item.

There are multiple ways to issue each of the three Clipboard commands. Table 4-3 summarizes them. Not every method is available in every application; the most universal method is the keyboard method.

TABLE 4-3 **Clipboard Actions**

Cut	Copy	Paste
Press Ctrl+X on the keyboard	Press Ctrl+C on the keyboard	Press Ctrl+V on the keyboard
Cut command on a menu (Usually on the Edit menu)	Copy command on a menu (Usually on the Edit menu)	Paste command on a menu (Usually on the Edit menu)
Cut icon on a toolbar or Ribbon	Copy icon on a toolbar or Ribbon	Paste icon on a toolbar or Ribbon
Right-click the selection ➪ Cut	Right-click the selection ➪ Copy	Right-click the selection ➪ Paste

An example will make the Clipboard concept clearer. Follow these steps if you want some hands-on practice with it. If you're already comfortable with the Clipboard, skip these steps:

1. **Start the Notepad app.**

   To do this, click the Start button, start typing Note, and then click Notepad in the search results.

2. **Start the WordPad app.**

   To do this, click the Start button, start typing WordPad, and then click WordPad.

3. **Switch to the Notepad window.**

   To do this, click the Notepad button on the taskbar.

4. **Type your name.**

5. **Select your name.**

   There are several ways you can do this. Here are two methods:

   - Drag over it with the mouse

   - Click where you want the selection to begin and hold down the Shift key as you press an arrow key to extend the selection.

   You can tell that your name is selected because its background changes color.

6. **Copy your name to the Clipboard using any method you like.**

For example, you could press Ctrl+C, or you could click Edit on the menu bar and then click Copy, as shown in Figure 4-19.

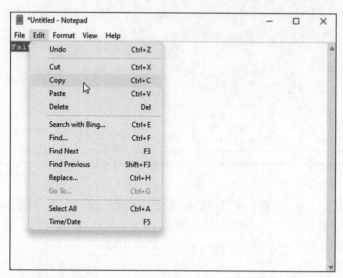

FIGURE 4-19

7. **Switch to the WordPad window.**

To do this, click the WordPad button on the taskbar, or click any visible portion of the WordPad window.

8. **In WordPad, type** Hello, my name is. **Then press the spacebar once.**

The insertion point appears after the space you just typed. The insertion point is a flashing vertical line that indicates where the next typed characters will appear — or, in this case, where whatever you paste from the Clipboard will appear.

9. **Paste your name from the Clipboard using any method you like.**

For example, you could press Ctrl+V, or click the Paste button on the Home tab of the Ribbon, as shown in Figure 4-20.

10. **Close both apps.**

Use any method you like for closing the apps. For example, you can click the Close (X) button in the upper-right corner of each window. Do not save your changes when prompted.

DRAG AND DROP

There is also a non-Clipboard method of moving and copying data between apps. You can select something and then drag it to the destination using the mouse. This can be a tricky method if you're not skilled with the mouse, and it's sometimes uncertain whether you'll end up with a Move or a Copy operation when you drag. (Let's just say "it's complicated" for the moment.) If you want to make sure you get a move, hold down Shift as you drag. If you want to make sure you get a copy, hold down Ctrl as you drag.

Paste

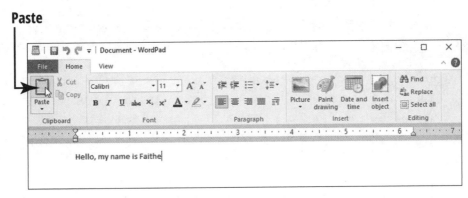

FIGURE 4-20

Install New Apps

You will probably want additional apps to supplement the basic set that come with Windows. Some are free; some are not. For the non-free ones, prices range from just a few dollars for a simple Microsoft Store app that tracks your exercise and eating habits to thousands of dollars for high-end professional video editing software. No matter what you want to do with your computer, chances are good that "there's an app for that."

Installing an app does three important things: It copies the app's files to your hard drive, it creates a shortcut to the app on the Start menu's list, and it modifies any system files as needed to allow the app to operate properly.

NOBODY RIDES FOR FREE

Most "free" apps are not completely free. The companies and people who make them have to get paid somehow. Here are some of the ways that happens:

Advertising. There are ads in the app, and you "pay" by watching them. The advertisers then pay the app owner. In some apps you can pay to get an ad-free version.

In-app purchases. The basic app is free to use, but there are extra features that you don't get unless you pay.

Free trials. The app is free to use for a certain amount of time, but then it expires unless you pay.

Popularity. Some companies make certain apps available for free as part of a larger strategy to make their products the most popular in a particular class. This is particularly true of web browser apps; they're all free because all the companies that make them are jockeying for position.

Remember earlier in the chapter when you learned that there are two kinds of apps: desktop and Microsoft Store apps? Each of those types has a different way of acquiring and installing them.

Installing a desktop app involves running a Setup utility that prompts you for information and installs the app on your computer.

To install a desktop app, follow these steps:

1. **Acquire the app, usually by downloading its Setup file.**

Desktop apps can come from various sources; the companies that create them sell them directly to customers, either through their own website or through an online retailer. You may also be able to buy copies of desktop apps on CDs or DVDs in local stores, or order CDs or DVDs online to be shipped to your home.

2. **Locate the Setup file in File Explorer and double-click it to start the setup.**

   The downloaded file is probably in your Downloads folder. Or, if you bought the app on a CD or DVD, you can view the disc's contents using File Explorer to locate the setup file. You learn how to use File Explorer in Chapter 6.

3. **Follow the prompts to complete the setup.**

   If you see a User Account Control dialog box asking if you want to allow this app to make changes to your device, click Yes.

4. **Start the new app from the Start menu.**

   Newly installed apps usually appear at the top of the Start menu.

Microsoft Store apps can only be acquired through Microsoft, although they are owned, updated, and supported by individual companies. In order for an app to be distributed through the Microsoft Store, the app must go through testing to demonstrate that it will run well under Windows, and it will not cause any harm to people's computers who install it.

Microsoft apps do not have a Setup utility; when you choose the app in the Microsoft Store, it is automatically downloaded and installed.

To install an app from the Microsoft Store, follow these steps:

1. **Open the Microsoft Store app in Windows.**

   To do this, click Start, type Store, and then click Microsoft Store.

2. **In the Store app, locate an app you want to install.**

   You can either browse the Store app or use its Search box to search by keyword for an app. If you aren't sure about an app, read online reviews of it before paying money for it. If you are just doing this for practice, choose a free app such as an entertainment app like Pandora or Spotify, or a free game like the one shown in Figure 4-21.

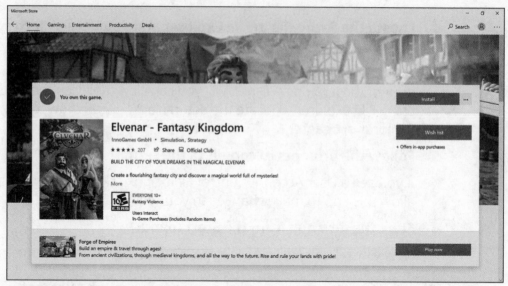

FIGURE 4-21

3. **On the app's page, click the Get button.**

   The app begins downloading and installing automatically. In most cases you should not have to do anything else to install it.

   When the app is fully installed, a Launch button (or sometimes a Play button if it's a game) appears.

4. **Click the Launch button (or Play button) to start the app.**

Remove Apps

If you decide you don't like a certain app, you can remove it from your PC. Removing an app is a tidy thing to do because it takes it off your Start menu, removes its files from your hard drive, and cleans up any pointers to it in your system files.

To remove a desktop app, follow these steps:

1. **Open the Settings app.**
2. **Click Apps.**

3. **(Windows 11 only) Click Apps & features.**

This isn't necessary in Windows 10 because step 2 takes you directly to the Apps & features screen.

4. **Scroll down to see the list of installed applications.**

The list is alphabetical. You can save some time if you want by typing the app you are seeking in the Search box above the list to narrow it down.

5. **Click the app you want to remove.**

In Windows 11, you must click the More button (the three vertical dots) to the right of the app's name. In Windows 10 you can click anywhere on the name.

In Windows 11, a menu appears with commands for managing the app, such as Modify and Uninstall. See Figure 4-22. In Windows 10, these options appear as buttons below the app's name.

Click this button to open the menu for that app

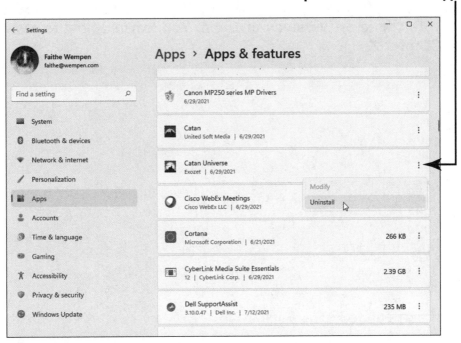

FIGURE 4-22

6. **Click Uninstall.**

   A confirmation prompt appears, containing another Uninstall button.

7. **Click the second Uninstall button.**

   A User Account Control dialog box appears asking if you want to allow this app to make changes to your device.

8. **Click Yes.**

   What you see at this point varies depending on the app. Each app has its own uninstall utility.

9. **Follow the prompts to complete the uninstall.**

   You may be prompted to restart your computer after the uninstall completes. Make sure you do so if prompted.

TIP

It's actually not a bad idea to restart your PC after an uninstall even if you are not prompted to do so. Many apps make changes to your system files when they install, and then they reverse those changes when you uninstall them (ideally, anyway). Restarting reloads your system files, making sure that you're working with the latest versions of them. You learned how to restart in Chapter 2.

Calculator app

» Writing brilliant documents with WordPad

» Jotting quick notes with Notepad

» Setting alarms and timers

» Keeping up on the weather

» Saving time with Cortana

» Exploring other Windows 10 apps

Chapter **5**

Six Great Apps that Come with Windows

Think about all the mundane tasks you do every day. Maybe you set your alarm clock to wake at the same time every morning, for example. You jot down a shopping list on the back of an envelope. You write a letter to a friend. Maybe you turn on the Weather Channel to see what it's going to be like outside, and then make another list of things you don't want to forget to do while you're out, like picking up the dry cleaning or getting your oil changed.

Your Windows computer can help you with a lot of those chores, and many others, too! In this chapter, you learn about six super-handy apps that come with Windows and how they can make your daily life easier. This chapter ends with a table listing other built-in apps in Windows and what they're good for.

Do the Math with the Calculator App

Remember when pocket calculators became affordable and everyone suddenly had one? They seemed like such a huge step forward in convenience, compared to doing calculations by hand or using a bulky adding machine. Today's calculators are much more sophisticated than those early ones we marveled at. They do scientific functions, and some of them even have video screens that display graphs of equations. They cost a pretty penny, too.

Windows has its own built-in Calculator app that can mimic the functionality of several kinds of physical calculators. With the Calculator app, you never have to wonder which closet you stashed your physical calculator in, or whether it's the right kind to do the math you need to do.

Start the Calculator app the same way you start any other app in Windows: from the Start menu. You can choose Calculator from the alphabetical list of apps, or you can type the first few letters and then select Calculator from the search results.

The default calculator view is Standard, shown in Figure 5-1. It's your basic calculator, with the four standard math functions (addition, subtraction, multiplication, and division) plus a few extras, like square roots and exponentiation.

To use it, you can click the buttons to enter numbers and choose math operators, or you can use your keyboard to enter the numbers and math symbols.

To switch to a different type of calculator, click the hamburger button (the button that looks like three horizontal stacked lines) to open a menu and then choose the type of calculator you want. There are five types of calculators available: Standard, Scientific, Graphing, Programmer, and Date Calculation. There are also two converters: Currency and Volume. Figure 5-2 shows the Scientific calculator.

Hamburger button (opens a menu)

FIGURE 5-1

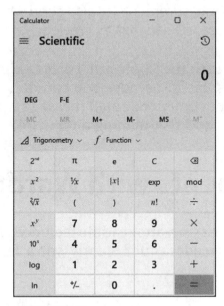

FIGURE 5-2

Most of these calculator types are self-explanatory to the people who need them. For example, if you're a scientist, you will understand all the buttons on the scientific one, and if you're an IT professional you'll understand how the programmer calculator converts between different numbering systems (decimal, octal, binary, and so on).

What may not be obvious, though, are these extras:

» **Date Calculation:** This calculator can tell you the difference between two dates, or can add or subtract days. For example, if you want to know how many days until Christmas, this calculator can tell you.

» **Currency Converter:** This calculator pulls the latest currency exchange rates from the Internet and converts between any two supported currencies.

» **Volume Converter:** This calculator converts between any two volume measurements, such as between milliliters and teaspoons. Very handy if you're trying to translate something between countries with different systems of measurement! It's also good when you want to increase or decrease the servings in a recipe.

TIP

To copy the result of a calculation to the Clipboard, you can right-click it and choose Copy. That's useful because you can do some math and then copy it over to a document or email message without having to retype the answer and risk making a typo.

Write Brilliant Documents with WordPad

WordPad is a simple word processing application included in Windows 10. It's been around for decades, but recently got a facelift to use a Ribbon for its interface.

The main selling point of WordPad is that it's free and already installed. It lacks a lot of the powerful features that you would expect from a full-featured word processor like Microsoft Word, but if you're just writing some simple correspondence, you don't really need fancy features.

ALTERNATIVES TO WordPad

If you find that WordPad isn't full-featured enough for you, there are several other attractive options. If you have Internet access, check out Word Online, a streamlined version of Microsoft Word available at www.office.com. You sign in with your Microsoft account at Office.com to use Word Online for free, and you can easily access your OneDrive online storage area from there, too. (More about OneDrive in Chapter 12.) The main drawback is that it's only available when you're online; if your Internet goes out, you can't use it. Google Docs is another similar free online-only word processor (docs.google.com).

If Word Online isn't full-featured enough either, consider the Microsoft Office suite of applications, which includes the full Microsoft Word app. You can get a Microsoft 365 yearly subscription for about $99, which gives you access to all of Microsoft's best full-featured business applications. There are links to it at Office.com.

Start WordPad from the Start menu, the same as other applications. It starts up with a new, blank document, and you can just start typing.

The Home tab of the Ribbon contains many formatting commands, arranged into groups. The group name appears below the commands, as you can see in Figure 5-3:

» **Clipboard:** To access the Windows Clipboard to cut, copy, or paste, use the commands in the Clipboard group.

» **Font:** To change the look of the lettering, use the commands in the Font group.

» **Paragraph:** To change the paragraph alignment and spacing, or to create bulleted or numbered lists, use the commands in the Paragraph group.

» **Insert:** To insert other content besides text, use the commands in the Insert group.

» **Editing:** To find or replace text, or to select all the text in the document at once, use the commands in the Editing group.

Ribbon

Quick Access Toolbar

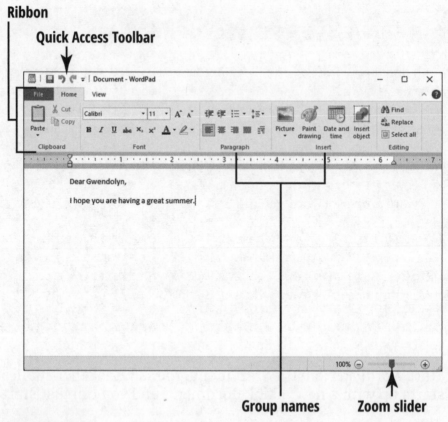

Group names　　**Zoom slider**

FIGURE 5-3

On the View tab of the Ribbon are commands for controlling how your document appears onscreen. You can zoom in and out, display or hide the ruler or status bar, and other viewing basics. You can also zoom in and out by dragging the Zoom slider in the lower right corner of the WordPad window.

TIP

The tiny toolbar above the Home tab on the Ribbon is a Quick Access Toolbar (QAT). It contains shortcuts to a few popular commands. By default there are three: Save (which saves your work), Undo (which reverses your last action), and Redo (which reverses your last reversal). You can customize what appears on the QAT by clicking the down arrow to its right to open a menu, as shown in Figure 5-4. The commands with checkmarks next to them already appear on the QAT.

FIGURE 5-4

Click the File tab to open a menu containing some essential commands for saving and printing your work, as shown in Figure 5-5. Notice that when you point the mouse at a command with an arrow to its right — namely Save As and Print — a panel appears to the right with additional options.

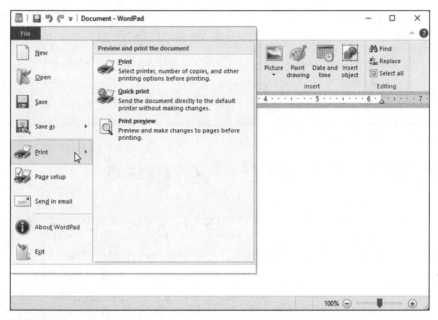

FIGURE 5-5

Table 5-1 provides a guide to the commands on the File menu.

TABLE 5-1   **Commands on the File Menu in WordPad**

Command	Purpose
New	Starts a new document.
Open	Opens a dialog box for choosing a saved file to reopen.
Save	Opens a dialog box for saving your document if it hasn't been saved before; otherwise it silently re-saves the changes to the current document.
Save As	Opens a dialog box for saving the current document with different settings than when it was saved before (such as a different name or location)
Print	Prints the document (after asking you to specify some printing options)
Page Setup	Opens a dialog box for setting the page size, page orientation, margins, and for selecting whether page numbers appear on the printed pages.
Send in Email	Opens your default email application and starts a new email message with the current document as an attachment.
About WordPad	Displays information about the WordPad application and about Windows.
Exit	Exits the WordPad application.

Jot Quick Notes with Notepad

WordPad is great when you need to create some nicely formatted text, but what if you don't care about formatting it? If you just want plain text, then Notepad may be a better choice for your project *du jour*. I use Notepad all the time to write my shopping and to-do lists, for instance, because I can keep a little Notepad window open on my Windows desktop as I do other things and then hop over to it to add items as I think of them.

Start Notepad from the Start menu, and you see a plain blank window. Just start typing in it. There aren't many options to fuss with.

TIP

As you type, if your text doesn't wrap automatically to the next line when it reaches the edge of the window, open the Format menu and click Word Wrap to turn that feature on.

The File menu has commands for all the essential file functions: saving, opening, and printing. See Figure 5-6. These work the same as in WordPad — and in most other applications too. Those three operations are pretty standard across applications.

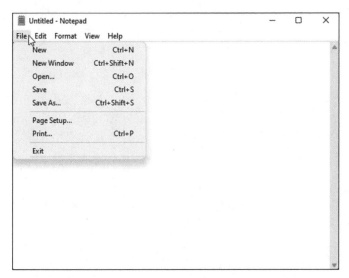

FIGURE 5-6

The Page Setup command on the File menu opens a Page Setup dialog box, where you can choose a paper size, page orientation, margins, and header and footer text. See Figure 5-7. The latter is text that you can set up to appear on every page of your printout at the top (header) and the bottom (footer). Click the Input Values hyperlink at the bottom of this dialog box to open a web page that explains the codes you can use in the header and footer.

WHY DOES NOTEPAD HAVE A FONT COMMAND?

On the Format menu there's a Font command that opens a Font dialog box, where you can choose a typeface, style, and size for your text. But wait a minute — I said earlier that Notepad is plain text only, didn't I? And it is. What you're actually choosing in the Font dialog box is the default font that Notepad uses to display your text. Your selection applies to all the text in your document. You can make the whole document in italics, but you can't make only one word italic. Furthermore, Notepad remembers your font selection, and makes it the default for all new documents in the future.

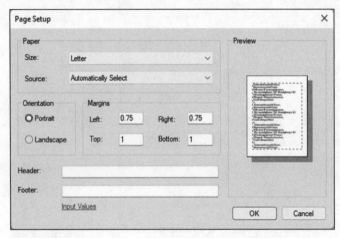

FIGURE 5-7

Set Alarms and Timers

Does anyone really know what time it is? You do, thanks to the built-in clock in Windows. It's always there for you, in the bottom right corner of the Windows desktop.

That clock is only the tip of the iceberg in Windows' time management tools, though. It also offers up timers, alarms, stopwatches, and even a world clock that tells you what time it is halfway across the globe.

To access all these timely tools (see what I did there?), open the Start menu and run the Alarms & Clock app. Then click one of the four services in the navigation bar on the left: Timer, Alarm, World Clock, or Stopwatch. Figure 5-8 shows the Alarm service ready to be set up, with two alarms already configured.

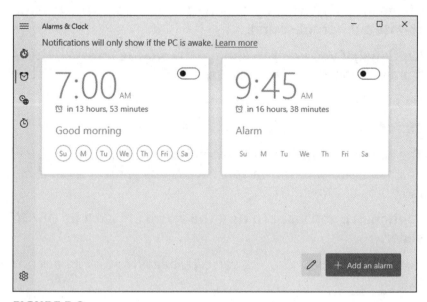

FIGURE 5-8

Just as a training exercise, let's set an alarm. Follow these steps:

1. Click Alarm in the navigation bar.

   The Alarm controls appear, as in Figure 5-8.

2. Click Add an Alarm.

   The Add New Alarm dialog box opens.

3. Set the desired alarm time by clicking the up or down arrows above the hour, minutes, and AM/PM boxes.

   For example, I am setting an alarm for 10:00 AM.

4. In the Alarm Name box, type a name for the alarm.

   I'm going to call this one Weekend Wakeup.

5. **(Optional) If you want to repeat the alarm, click the Repeat Alarm check box and then click the buttons for each day of the week that it should repeat on.**

For example, I'm going to repeat this alarm on Saturday and Sunday.

6. **(Optional) If you want a different alarm sound, open the Alarm Chime drop-down list by clicking the down arrow on it and choose a different sound.**

7. **(Optional) If you want to change the snooze interval, choose it from the Snooze Time drop-down list.**

Figure 5-9 shows a new alarm, ready to be saved.

8. **Click Save.**

The new alarm is saved, and appears on your list of alarms. It is enabled by default; you can tell because the On/Off slider is blue, indicating an On status.

9. **(Optional) If you want to turn the alarm off, click its On/Off slider.**

Its status toggles to Off. You can click it again to turn it back on if you like.

10. **Click the alarm (anywhere except on the On/Off slider).**

An Edit Alarm dialog box opens. It has the same settings as in Figure 5-9; the only thing different is the name of the dialog box.

11. **Make any changes to the alarm settings, and then click Save.**

Or, to delete the alarm, click the Delete icon (the red trash can) in the upper right corner of the dialog box.

TIP

As you work with alarms more and more, you'll probably accumulate several, and you might forget to delete them when they've served their purpose. To delete several alarms at once, click the Edit Alarms icon in the lower right corner of the Alarms & Clock window and then click the red trash can icon on each of the alarms that you want to delete.

Alarm name

Days to repeat the alarm

Alarm chime sound

Snooze interval

FIGURE 5-9

I showed you the Alarm section of the app because it's the most complex of them; you can explore the Timer, World Clock, and Stopwatch features at your leisure.

Keep Up on the Weather

Whether you're a weather enthusiast or you just want to know if you should take an umbrella when you run your errands tomorrow, you'll appreciate the Weather app in Windows.

Like many of the other apps you've seen so far, the Weather app has a navigation bar along the left side of its window. You can click the buttons on it to move to different sections of the app. Click the hamburger button to open a pane that shows the names of the buttons. Figure 5-10 names them for you, and shows the screen you get when you choose Forecast.

FIGURE 5-10

In the Maps section, you can view a variety of weather maps, including Temperature, Radar Observation, Radar Forecast, Precipitation, Satellite, and Cloud. You can change the region via the Change Region drop-down list in the upper right corner of the map, and you can zoom in and out with the + and – buttons. See Figure 5-11.

FIGURE 5-11

In the News section is a set of links to news stories about weather, constantly updated from online news services.

The Historical Weather section shows weather statistics over time in your area. You can choose Temperature, Rainfall, or Snow Days.

In the Favorites section, you can set up additional locations for which you want to monitor the weather. I like to set up the locations where my family members live so I can see what kind of weather each one is having each day.

The Settings icon (which looks like a cog) in the bottom-left corner enables you to customize the way the Weather app looks and operates. For example, you can show the temperature in Fahrenheit or Celsius, and you can set a launch location (which is the location that appears in the Forecast when you start up the app).

Save Time with Cortana

Cortana is a personal assistant feature that's part of Windows. Cortana resembles apps on smartphones such as Siri on iPhones that reply to your voice requests for directions, a list of nearby restaurants, current weather conditions, and more.

Cortana can also interact with other apps. For example, you can ask Cortana to tell you the weather forecast (Weather app), to set an alarm or timer (Alarms & Clock app), and to create an appointment on your calendar (Calendar app).

In Windows, to try out Cortana for yourself, follow these steps.

1. **Open Cortana.**

   In Windows 10 you can click the Cortana icon on the taskbar. It's the circle to the right of the Search box.

   In both Windows 10 and Windows 11, you can open Cortana from the Start menu.

Cortana opens.

If you have not used Cortana before, you may see a Sign in to Cortana prompt. Follow these steps to sign in, if needed.

2. Type what you want and press Enter.

Or

Click the Speak to Cortana icon (which looks like a microphone) and then talk aloud to Cortana.

Figure 5-12 shows a conversation I had with Cortana about tomorrow's weather, for example.

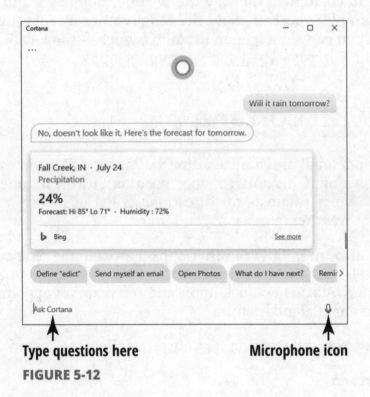

Type questions here   **Microphone icon**

FIGURE 5-12

What can you ask Cortana to do? Just about anything that you could do yourself using Windows! Here is just a sampling of things you could say (or type):

» Set an alarm for 8 a.m.

» Open Notepad.

» Will it rain tomorrow?

» Who was the 42nd president of the United States?

» What is 42 times 16?

» Define *paripatetic*.

» Find pizza restaurants nearby.

» Tell me a joke.

TIP

You can set up Cortana to hear you giving commands even when the Cortana window isn't open. To do that, and also explore other settings you can adjust, follow these steps:

1. **In Cortana, click the Open Menu button (. . .) in the upper-left corner.**

A menu opens.

2. **On the menu, click Settings.**

Under the This Device heading are three categories of settings you can adjust.

3. **Click Voice Activation.**

The Voice Activation settings appear. If you see the message All required permissions are turned on, you are all set. Otherwise, follow the instructions that appear to turn on voice activation.

4. **When you are finished with voice activation, click the Back button (left-pointing arrow) to return to the Settings screen. Then, click Back again to return to Cortana.**

After enabling voice activation, you can wake up Cortana by saying "Cortana" at any time; you don't have to click the Cortana icon or the microphone in the Cortana app window.

Explore Other Windows Apps

A full explanation of all the Windows apps would take an entire book, so I just touched on a handful of the best ones in this chapter. But there are plenty more that you may want to explore on your own. Table 5-2 lists them. To use one, just click the Start button and then start typing the app's name; then click the app in the search results. Some of these apps are covered in detail later in the book; where that's the case, I've included the chapter reference in the table.

TABLE 5-2   **Other Apps That Come with Windows**

App	What It Does
Calendar	Keeps track of your appointments and reminds you when they are coming up soon.
Camera	Enables you to access your computer's camera to take still photos and record videos.
File Explorer	Manages files and folders. See Chapter 6.
Get Help	Provides tutorials and other help for using Windows.
Mail	Enables you to send and receive email. See Chapter 11.
Maps	Helps you find locations and get directions to them.
Movies & TV	Enables you to play personal video clips and access online clips from other sources. See Chapter 15.
OneNote	Organizes links, notes, graphics, and other material you might collect for a project.
Paint	Creates original artwork using drawing and painting tools.
Paint 3D	Creates 3D artwork.
People	Organizes your list of contacts, like in an address book. See Chapter 11.
Photos	Organizes your saved photos. See Chapter 15.
Skype	Enables you to place and receive video calls and participate in audio and video meetings. See Chapter 13.

App	What It Does
Snip & Sketch	Captures images of the Windows screen (called screenshots) and saves them as graphics.
Sticky Notes	Enables you to create your own digital sticky notes and place them on your desktop for easy access.
Teams (Windows 11)	Enables you to create groups of people and communicate with them to chat, share files, and more. You can download a free copy of Teams for Windows 10, but it comes automatically with Windows 11.
Video Editor	Enables you to create and save video projects that add effects, narration, and captions to your home videos.
Voice Recorder	Records audio narration using your PC's microphone. See Chapter 15.
Windows Fax & Scan	Enables you to send and receive faxes (if you have a fax modem in your PC) and to operate a scanner (if you have one). This app doesn't appear on the Start menu's app list in Windows 11 but you can find it by searching for it.
Windows Media Player	Enables you to play music from your private music collection and acquire new music, plus burn audio CDs and rip (copy) music from audio CDs to your PC. See Chapter 16.

organizes data

» Checking out the File Explorer interface

» Displaying different locations

» Finding files and folders

» Viewing file listings in different ways

» Selecting multiple items at once

» Moving, copying, deleting, and renaming files and folders

» Creating a shortcut to a file or folder

» Creating a compressed file or folder

» Adding a folder to your Quick Access list

» Backing up files

Chapter **6**

Managing Your Personal Files

oin me for a moment in the office of yesteryear. Notice all the metal filing cabinets and manila file folders holding paper rather than the sleek computer workstations, ubiquitous tablets, and wireless Internet connections we use today.

Fast forward: You still organize the work you do every day in files and folders, but today the metal and cardboard have given way to electronic bits and bytes. Files are the individual documents that you save from within applications, such as Word and Excel, and you use folders and subfolders in Windows File Explorer to organize files into groups or categories, such as by project or by year.

In this chapter, you find out how to organize and work with files and folders, including

» **Finding your way around files and folders:** This includes tasks such as locating and opening files and folders using various search tools, and using some of the tools in File Explorer.

» **Manipulating files and folders:** These tasks cover moving, renaming, deleting, and printing a file.

» **Squeezing a file's contents:** This involves creating a compressed folder to reduce the size of a large file or a set of files to be more manageable when backing up or emailing them.

» **Backing up files and folders:** To avoid losing valuable data, you should know how to make backup copies of your files and folders on a recordable CD/DVD or **flash drive** (a small stick-shaped storage device that fits into a USB port on your computer).

Understand How Windows Organizes Data

When you work in a software program, such as a word processor, you save your document as a file. Files can be saved to your computer hard drive or to storage media you can remove from your computer, such as USB flash drives. Note that you can also save files to an online storage site, such as OneDrive; this is known as **storing in the cloud**. Working with OneDrive and the cloud is covered in detail in Chapter 12.

REMEMBER

You can organize files by placing them in folders that you work with in an app called File Explorer, which is included with Windows. The Windows operating system helps you organize files and folders in the following ways:

» **Take advantage of predefined folders.** Windows sets up some folders for each user account automatically, such as Documents, Downloads, Music, Pictures, and Videos (see Figure 6-1). Certain apps use these folders as default locations for their content. For example, most text-based apps use the Documents folder by default.

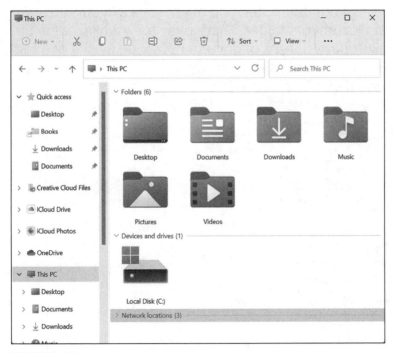

FIGURE 6-1

» **Create your own folders.** You can create any number of folders and give them names that identify the types of files you'll store there. For example, you might create a folder called *Digital Scrapbook* if you use your computer to create scrapbooks, or a folder called *Taxes* where you save emailed receipts for purchases and electronic tax-filing information.

» **Place folders within folders to further organize files.** A folder you place within another folder is called a **subfolder**. For example, in your Documents folder, you might have a subfolder called Retirement in which you store information about all your retirement activities. Then, within the Retirement folder, you might have subfolders for Vacations, Remodeling, Hobbies, and Finances, as shown in Figure 6-2.

» **Move files and folders from one place to another.** Being able to move files and folders helps when you decide it's time to reorganize information on your computer. For example, when you start using your computer, you might save all your documents to your Documents folder. That's okay for a while, but in time, you might have dozens of documents saved in that one folder. To make your files easier to locate, you can create subfolders by topic and move files into them.

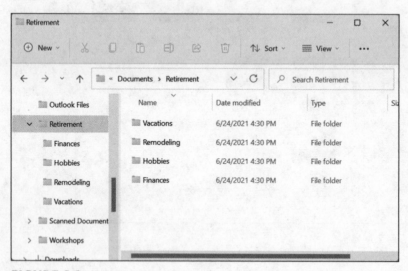

FIGURE 6-2

TECHNICAL STUFF

The folders you will work with are primarily those in your own user account folder. Each user account in Windows has its own personal folder set, to give people who share a computer some privacy. Your folders may appear in File Explorer to be at the top level of things — and that's good, because they're handy that way! But they are actually stored in a location called C:\Users*yourname* where *yourname* is your user name (or an abbreviated version of it).

WARNING

System files are stored in the C:\Windows folder. Never delete or rename any files in this folder! The same goes for the folders where application files are stored: C:\Program Files and C:\Program Files (x86).

WHAT IS ONEDRIVE?

Besides the files on your local PC, you might interact with files in OneDrive using File Explorer. You'll learn all about OneDrive in Chapter 12, but for now, just know that it is a cloud-based storage system that Microsoft provides to all Microsoft account holders. It provides a secure way of storing your personal files so that they are available from any computer and any location, and you never need to worry about backing them up because Microsoft's servers will always have your back.

I bring up OneDrive now for a couple of reasons. One is that you'll see OneDrive as a location in File Manager. You can save and retrieve files on OneDrive as easily as you can on your local PC. So, if you want to start using OneDrive to store your personal files, that's fine. Go for it.

The other reason is that your OneDrive has some default-created folders that have the same names as your local personal folders, including Documents and Pictures. It can be easy to get the local and OneDrive versions of these files confused — you might think you have stored a file in one place but it's actually in the other. Knowing this can help you track down a misplaced file without panicking.

Explore the File Explorer Interface

To start File Explorer, click the File Explorer icon on the taskbar. When File Explorer opens, it displays the Quick Access location at first. From there you can navigate to any other location (which you'll learn in the next section).

Before you start using File Explorer, though, you may want a quick tour of the interface so you'll understand the instructions coming up in the rest of the chapter. Figures 6-3 and 6-4 show Windows 11 and Windows 10, respectively. Have a look at these figures as I take you on the following tour.

REMEMBER

The File Explorer app window consists of these parts:

>> A toolbar at the top. In Windows 10, it's a ribbon-style toolbar, like you learned about in Chapter 4. In Windows 11, it's a much simpler affair with a single row of buttons. In the Windows 11 version, a More button (. . .) provides access to additional commands.

>> A navigation pane on the left side. You can quickly click other locations to jump to them at any time.

>> A Quick Access section at the top of the navigation pane. You can put shortcuts here to the locations you use most often.

>> An Address bar below the toolbar. This bar shows the current location you are seeing. In Figure 6-3, that's Documents ⇨ Retirement.

>> A Files pane on the right side. This is where the content of the current location appears. In Figures 6-3 and 6-4, the Files pane shows four subfolders. If the Retirement folder contained any files, they would appear here too.

TIP

Notice in Figure 6-3 that the toolbar in Windows 11's File Explorer is a bit lacking on labels compared to Windows 10's much more robust Ribbon. It just shows a bunch of icons. How do you know what they do? Easy! Just point at an icon with the mouse pointer and you'll see a pop-up ScreenTip telling you its name. Do this now for each of the icons on the toolbar, to familiarize yourself with what's there. You'll need those icons later in this chapter.

Toolbar   **Address bar**

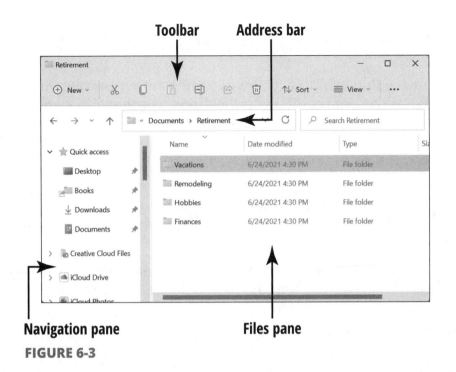

Navigation pane   **Files pane**

FIGURE 6-3

Ribbon   **Address bar**

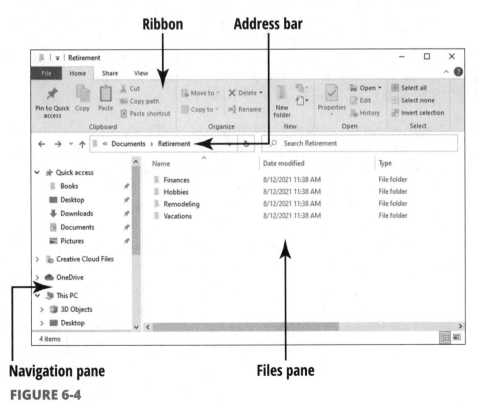

Navigation pane   **Files pane**

FIGURE 6-4

Move between Different Locations

As you use File Explorer, you will want to move between locations — in other words, make different locations' content show up in the Files pane. Here are some basic tips for moving around:

» To open a subfolder that you see in the Files pane, double-click it.

» To go back to the last location you looked at, click the Back arrow (the left-pointing arrow) to the left of the Address bar. A keyboard shortcut is Alt+left arrow.

» To return to where you were before you used the Back arrow, click the Forward arrow (the right-pointing arrow). A keyboard shortcut is Alt+left arrow.

» To go up one level in the folder hierarchy, click the Up arrow.

You can also choose a location from the Address bar. The best way to understand this option is to try it out. In the Address bar, notice the little right-pointing arrows between each part of the path. Click one of those arrows to open a menu of other locations at that level of the hierarchy, and then click one of the locations on that menu to jump to it. See Figure 6-5.

Another way you can get around is to click a location in the navigation bar. The navigation bar contains a collapsible tree-like structure of all the available locations. The basic top levels are:

Quick Access: Shortcuts to your favorite locations; you can customize this list, as you'll learn later.

OneDrive: Access to your Microsoft-provided cloud storage system. More on this in Chapter 12.

This PC: Access to the storage devices on your local computer, such as your hard drive and any other drives that might be connected. Under this category are also shortcuts to commonly used locations like Desktop, Documents, Downloads, Music, Pictures, and Videos.

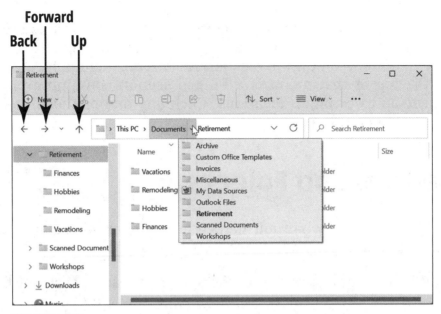

FIGURE 6-5

> **Network:** Access to any drives that are shared on your local area network, if you have one.

Depending on what other software is installed on your computer, there may be other top-level items in the navigation bar too. If you look back at Figure 6-1, you'll notice two such items: Create Cloud Files and iCloud Photos.

Notice that these top-level locations have little arrows to their left. If the arrow is pointing to the right, the level is collapsed. If the arrow is pointing down, the level is expanded. Click the arrow to switch between the two states.

Each subordinate level that has subfolders within it has its own little arrow that you can use to collapse or expand it. Expanding a level makes the subfolders beneath it visible in the navigation pane; when you see the location you want to jump to, you can click it to display its content in the Files pane.

TIP

Here's a trick for opening a specific location in File Explorer at the same time as you are starting File Explorer. Right-click the File Explorer icon on the taskbar. A shortcut menu appears, and at the top of that menu is a Pinned section containing pinned shortcuts to common locations. Click any one of those to start File Explorer at that location.

Locate Files and Folders

Can't remember what you named a folder, or where you saved it? There are multiple ways to solve that problem.

In Windows 11, you can open the Start menu and check out the Recommended section at the bottom. The files you've most recently accessed appear here, as in Figure 6-6. You can click one of them to reopen it in its native app. Only a couple of files appear here by default, but if you click More, a much longer list appears to choose from. (This doesn't work in Windows 10.)

You can also use the Search feature to find it.

1. **If you have a general idea of the main section that the file is in, display that location.**

    For example, if you know it's on your local PC, navigate to This PC. Doing this saves time by searching only the areas where it might be.

2. **Click in the Search box in the upper-right corner, and begin typing the name of a file or folder.**

    Search results begin appearing immediately on a menu. See Figure 6-7.

3. **Click a file in the search results to open it.**

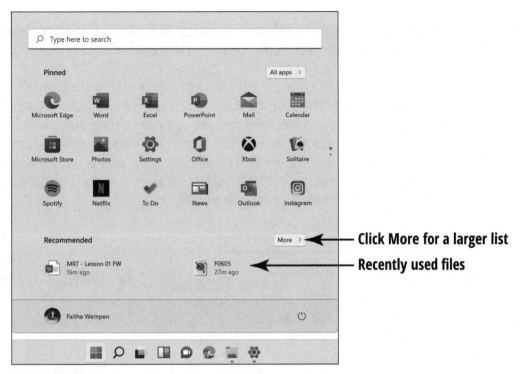

Click More for a larger list

Recently used files

FIGURE 6-6

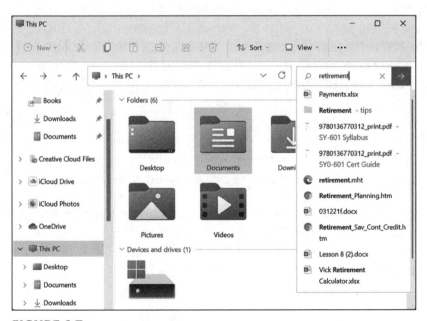

FIGURE 6-7

The method you just learned is quick and easy, but there's an alternate method that gives you more options. It's different between Windows versions, though:

» **Windows 10:** After typing the search term in step 2, press Enter. A full search runs, and the results appear in the Files pane (after a brief delay; be patient). From there you can double-click one of the found files to open it. This also has the side effect of opening a Search tab on the Ribbon, which you can use to fine-tune the search parameters. See Figure 6-8.

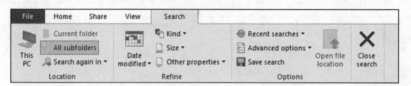

FIGURE 6-8

» **Windows 11:** You can do the same thing as in Windows 10 — press Enter at step 2 — and you'll get the results in the Files pane, the same as with Windows 10. However, you don't get the Search tools on the toolbar. Instead, click the More button (. . .) and click Search Options on the menu that appears; this menu, shown in Figure 6-9, contains the same commands and tools as on the Search tab on the Ribbon in Windows 10.

You can also search using Cortana. In Windows 11, you'll need to start the Cortana app first; in Windows 10 you can click the Cortana icon on the taskbar. Then, either type or say what you want Cortana to find.

And here's one final way to search: You can use Windows' own Search tool. In Windows 11, click the Search icon on the taskbar and then type what you are looking for. In Windows 10, click in the Search box on the taskbar and type directly into that box.

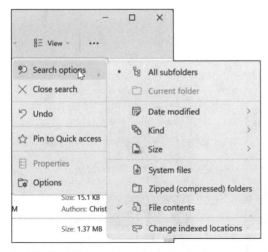

FIGURE 6-9

View File Listings in Different Ways

File Explorer provides a variety of viewing choices for the lists of files and folders that appear in the Files pane.

For starters, you can change the icon size, arrange the list differently, and show more or fewer details about each item. To make this change:

» **Windows 11:** Click View on the toolbar and then choose View from the menu that appears.

» **Windows 10:** Click the View tab and then make a selection in the Layout group.

TIP

The larger icon sizes commonly show previews of certain data file types as thumbnail images on them, so they can be useful when browsing folders that contain pictures or documents. Figure 6-10 shows a folder containing photos in Extra Large view, for example. Figure 6-11 shows the same folder in Details view, where there's no preview of the image but plenty of information about each file.

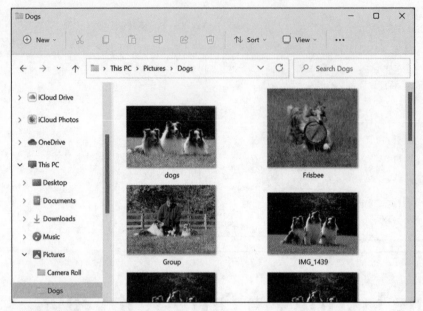

FIGURE 6-10

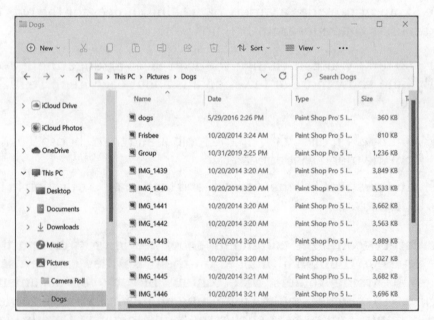

FIGURE 6-11

If you want to preview a data file without changing the view, you can turn on the Preview feature. This opens a preview pane to the right of the Files pane, and it previews whatever file is selected. (It can only preview one file at a time.)

To access the Preview pane:

» **Windows 11:** Click the View button, point to Show, and click Preview pane.

» **Windows 10:** On the View tab, click the Preview pane button.

The Details pane is similar to the Preview pane; it's an extra pane that shows details about the selected file or folder, similar to the details that appear in Details view. The Details pane and the Preview pane are mutually exclusive; when you turn one on, the other goes off. (To turn them both off, select the command again for the one that's currently on.)

To access the Details pane:

» **Windows 11:** Click the View button, point to Show, and click Details pane.

» **Windows 10:** On the View tab, click the Details pane button.

You can also sort the file listing according to different properties such as file name, file type, size, or date. In Details view, it's easy to sort: just clicking the column heading by which to sort. Click it again to reverse the order.

Sorting is a little harder (but still doable) in other views.

1. **In Windows 11, click the Sort button on the toolbar.**

   OR

   In Windows 10, click the Sort button on the View tab.

2. **On the menu that appears, click the property you want to sort by.**

The list in Windows 11 contains only a few properties but you can click More to see a longer list.

The initial list in Windows 10 contains all the available properties.

3. **(Optional) If you want to reverse the order, click the Sort (or Sort by) button again and then click Ascending or Descending.**

Select Multiple Items at Once

You can act on multiple items at once. And by "act on," I mean do some of the things that are coming up next in this chapter: moving, copying, and deleting. To affect multiple items, though, you must select them before you issue the command.

To select multiple contiguous items (that is, items that are next to each other on the list), click the first one and then hold down the Shift key as you drag across the others. Release the mouse button when all the items you want are selected.

To select multiple non-contiguous items, click the first one and then hold down the Ctrl key as you click each additional item individually. Release the Ctrl key when you're done.

Move or Copy an Item

Sometimes you might save a file or folder in one place but then decide you want to move the item to another location. You can move it there, or you can make a copy in the new location, so it's in both places.

There are lots of methods for moving and copying in File Explorer, but frankly, many of them are more trouble than they're worth. So I focus here on two methods that are quick and work well. Depending on the situation, you may find yourself switching between them.

First, in File Explorer, navigate to the location where the file or folder is stored, and then select it by clicking it in the Files pane. Then do any of the following:

» **Use the Clipboard.** You can use the Windows Clipboard to move an item with a combination of the Cut command and the Paste command. Similarly, you can copy an item with a combination of the Copy command and the Paste command. The first command (Cut or Copy) places the item on the Clipboard, a temporary holding area, and then the second command (Paste) places whatever is on the Clipboard at the active location.

For example, here's how a copy operation works. Select the item, and then issue the Copy command. See Table 6-1 for a list of the different options for doing that. (The keyboard method is my go-to.) Then in File Explorer, navigate to the destination location, and then issue the Paste command.

TABLE 6-1   Clipboard Actions

	Copy	Cut	Paste
Keyboard method	Ctrl+C	Ctrl+X	Ctrl+V
Toolbar method (on Home tab in Windows 10)	Copy button	Cut button	Paste button
Right-click method	Right-click the item and choose Copy	Right-click the item and choose Cut	Right-click the item and choose Paste

» **Drag-and-drop the item where you want it.** To do this, the destination location must be visible, either in the Files pane or in the navigation pane. You might need to expand some levels in the navigation pane to bring the destination location's name into view. Then point to the item with your mouse pointer, and hold down the left mouse button as you move the mouse to the destination. When you reach the destination, release the mouse button.

MOVE OR COPY?

Here's something that's potentially confusing. Sometimes drag-and-drop moves things, and sometimes it copies things. It all depends on the relationship between the original location and the destination location. Drag-and-drop moves items when the item and its destination are on the same volume (that is, the same drive, like your hard drive). When the item and its destination are on different drives, drag-and-drop creates a copy rather than moving. You can force it to move by holding down the Shift key as you drag, and you can force a copy by holding down Ctrl as you drag. Some people just get into the habit of always holding down the key for what they want so they don't have to think about it.

You can also drag-and-drop between two File Explorer windows. To open a second File Explorer window, right-click the File Explorer icon on the taskbar and choose File Explorer; a new window opens. You can also right-click any folder and choose Open in new window.

Delete or Undelete an Item

If you don't need a file or folder anymore, you can clear up clutter on your computer by deleting it. As with most other activities in File Explorer, there are multiple ways to do it. You pick which one you like the best in each situation.

Select the file or folder that you want to delete and then do any of the following:

>> Press the Del key on the keyboard.

>> Right-click the item and choose Delete.

>> Drag the item to the Recycle Bin icon on the desktop.

>> Click the Delete button on the toolbar (Windows 11) or on the Home tab on the Ribbon (Windows 10).

TIP

When you delete a file or folder in Windows, it's not really gone — at least not right away. It's moved to the Recycle Bin, a special folder that stores items slated for deletion. There is a shortcut to it on the desktop. Windows purges deleted items from this folder as needed to reclaim space on the hard drive, but you may still be able to retrieve recently deleted files and folders from it.

To restore a deleted file or folder, double-click the Recycle Bin icon on the desktop to open its window. Right-click the file or folder and choose Restore. Windows restores the file to wherever it was before you deleted it.

To empty the Recycle Bin so that nobody can restore your deleted items, right-click the Recycle Bin icon on the desktop and click Empty Recycle Bin.

Rename an Item

You may want to change the name of a file or folder to update it or differentiate it from other files or folders. Select the item in File Explorer, and then do any of the following:

» Click the Rename button on the toolbar (Windows 11) or on the Home tab (Windows 10). Type the new name and press Enter.

» Right-click the item and choose Rename. Type the new name and press Enter.

» Click the item's name, pause a second, and then click it again. This makes the name editable. Edit the name and press Enter.

» Press the F2 key on the keyboard. This makes the name editable. Edit the name and press Enter.

Here are some hazards to watch out for when renaming files.

» You should never rename Windows system files or files needed to run applications. Applications rely on a complex interconnected system of files to work properly, and renaming even one of them may break the app. Rename only your own personal data files.

» You should never change a file's extension. An **extension** is a code that appears after the file's name to indicate what kind of file it is. For example, a Word document named Travel might be Travel.docx. The *.docx* part tells Windows to open the file in Word. You don't usually see file extensions in File Explorer (they're hidden by default), so you don't have to worry about them. However, File Explorer does have a setting that allows you to see them (View ⇨ Show ⇨ File name extensions in Windows 11, or View ⇨ File extensions in Windows 10. And if that option is enabled, you *will* see the file extensions, and you'll need to be careful not to change them when renaming files.

» You can't rename a file and give it the same name as another file of the same type in the same location. Windows won't let you. However, if they are two different file types, they can have the same name but different extensions. If file extensions are hidden in File Explorer (see above), you won't know what their extensions are, so it might *seem* like two files have the same name. That's really confusing, right! To ward off that kind of chaos, try not to ever give two files the same name.

Create a Shortcut to an Item

You can place a shortcut to a file or folder you used recently on the desktop for quick and easy access. Do one of the following:

» Right-drag a file from File Explorer to the desktop. (In other words, drag using the right mouse button rather than the left.) When you release the mouse button, a menu appears. On that menu, click Create Shortcuts Here.

>> In File Explorer, right-click the item. In Windows 11, click Show More Options. (That step isn't needed in Windows 10). Then point to the Send To command and then select Desktop (Create shortcut). See Figure 6-12.

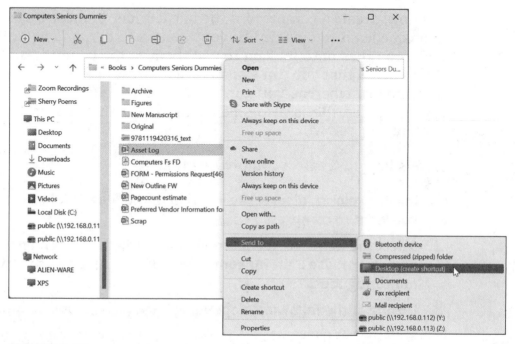

FIGURE 6-12

TIP

After you've placed a shortcut on the desktop, you can open the file in its native application or open a folder in File Explorer, simply by double-clicking the desktop shortcut icon.

REMEMBER

A shortcut icon is just a pointer to the original file; it's not a true copy of it. When you double-click a shortcut, Windows takes you to the original file that the shortcut refers to. When you delete a shortcut, the only thing you've deleted is the pointer to the original; the original file is still safe in its home location.

Create a Compressed File

To shrink the size of a file or all the files in a folder, you can compress them. This is often helpful when you're sending an item as an attachment to an email message.

Windows treats a compressed file like a folder in most ways. You can double-click it to open it like you would a folder, and you can drag-and-drop files onto its icon to add them to the compressed file. The difference is that you can transfer a compressed file as a single email attachment, whereas you can't transfer a regular folder via email. A compressed file is also smaller in size than the sum total of its content.

Here's how to create a compressed file:

1. **In File Explorer, display the location of the files or folders that you want to compress.**

2. **Select the items, as you learned to do earlier in this chapter. Remember: use Shift for contiguous selections, and use Ctrl for noncontiguous.**

3. **Do one of the following, depending on your Windows version:**

   - *Windows 11:* Right-click the selection and choose Compress to ZIP file. See Figure 6-13.

   - *Windows 10:* Right-click the selection, point to Send to, and choose Compressed (zipped) folder.

4. **Type a new name for the compressed file (replacing the default name, which is the same as the name of the last item in the selection). Then press Enter to accept the new name.**

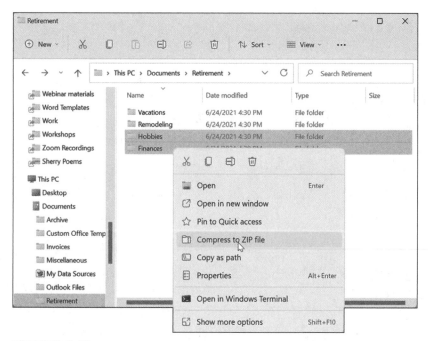

FIGURE 6-13

Customize the Quick Access List

Remember the Quick Access list in the navigation pane that you learned about at the beginning of this chapter? It offers a fast way to access frequently used folders. You can customize what's on that list so it shows the locations that you use the most. Never mind what other people think should be popular! You do you.

To customize the Quick Access list, open File Explorer and check out the current Quick Access list. Expand it if it's collapsed. Notice the little pushpin symbols next to each item. That means that each of those items is pinned there.

To add an item there, display it in the Files pane, select its icon, and then drag its icon and drop it onto the Quick Access list.

To remove an unwanted item, right-click it on the Quick Access list and choose Unpin from Quick Access. See Figure 6-14.

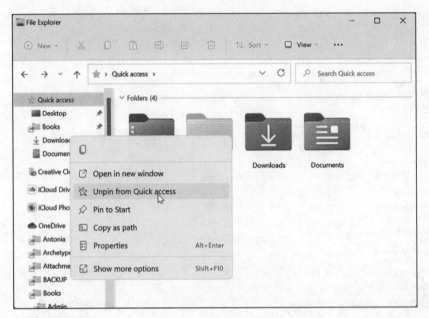

FIGURE 6-14

Back Up Files to an External Drive

You've put a lot of work into your files, so don't forget to back them up. If your computer is damaged or loses data, you'll then have a copy safely tucked away.

There are many different ways to back up your files. You can copy them to a USB flash drive or other external drive, for example. Here's how to do that:

1. **Connect a USB flash drive or other external drive to your PC.**

An icon for it should appear in the navigation pane, under the This PC heading.

2. **In File Explorer, navigate to the location containing the files to be backed up, and select them.**

3. **Copy the selection to the USB flash drive. Here are some of the ways you could do that:**

- Drag-and-drop the selection onto the USB flash drive's name in the navigation pane.

- Right-click the selection, choose Show More Options (Windows 11 only), point to Send to, and click the USB flash drive's name from the menu that appears.

- Open a second File Explorer window and display the USB flash drive's contents in it. Then drag-and-drop the files to the flash drive's window.

- Use the Copy command to copy the selection to the Clipboard. Then display the USB flash drive's contents in File Explorer's Files pane, and use the Paste command to paste them there.

TIP

You can also automatically save files to OneDrive, Microsoft's online file-sharing service, as discussed in Chapter 12. If you do that, you don't need to worry about keeping a local backup (although some people like to anyway, just to be doubly safe).

Start menu and taskbar

» **Customizing the Windows 10 Start menu and taskbar**

» **Changing the screen resolution and scale**

» **Applying a desktop theme**

» **Changing the desktop background image**

» **Changing the accent color**

» **Managing desktop icons**

» **Adding widgets to the desktop**

» **Making Windows more accessible**

Chapter **7**

Making Windows Your Own

Windows is pretty good right off the bat, but it gets even better when you add some customization to it. When you have your own pictures on the desktop, the right size font for your eyesight, and shortcut icons for all your favorite things, Windows will really start seeming like it's *yours*. And that's what this chapter is all about: tweaking the various defaults to bend Windows to your will. (Sounds kind of evil-genius-like when I say it that way, doesn't it?)

This chapter starts out by showing you how to customize the Start menu. The thing is, though, Windows 10 and Windows 11 are very different Start-menu-wise, so I break things out into separate sections here. You only need to read the section that applies to your version. Then we move on to customizations that affect the taskbar and the desktop. We finish up this chapter by looking at Ease of Access — in other words, all the ways that Windows can be adapted to help people with various types of physical challenges, like limited hearing, vision, and mobility. If that's not you yet, it might be someday, so it's good to be prepared.

Customize the Windows 11 Start Menu

You learned quite a bit about the various parts of the Start menu in previous chapters, so I assume you know the basics. Now let's see what we can do to make it better!

Follow these steps to check out the Start menu customization options:

1. **Right-click the desktop and choose Personalize.**

   This is a fast way to get into the Settings app's Personalization section.

2. **On the Personalization page of the Settings app, scroll down and click Start to open the Start menu options. See Figure 7-1.**

3. **Adjust any of the three options at the top of this screen to your preferences.**

   These options all pertain to various shortcuts appearing on the Start menu: Show Recently Added Apps (on by default), Show Most Used Apps (off by default), and Show Recently Opened Items in Start, Jump Lists, and File Explorer (on by default). The latter is the reason the Recommended section appears at the bottom of the Start menu. If you turn off this option, that goes away.

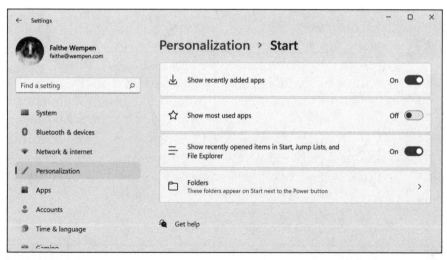

FIGURE 7-1

4. **Click Folders.**

A list appears of various folders you can add shortcuts for on the Start menu, next to the Power icon.

There aren't any by default, but you might like to enable Settings, for example, and some of the others too. See Figure 7-2.

5. **Close the Settings app.**

You can also customize the Start menu by changing the pinned apps:

» **Pin**: To pin an app to the Start menu that's not already there, click All Apps to locate the desired app; then right-click it and choose Pin to Start.

» **Move**: To move a pinned app to the top of the menu, right-click the app's pinned icon and choose Move to Top. You can also drag-and-drop a pinned icon to a different spot.

» **Unpin**: To unpin an app that's pinned there, right-click the app's pinned icon and choose Unpin from Start.

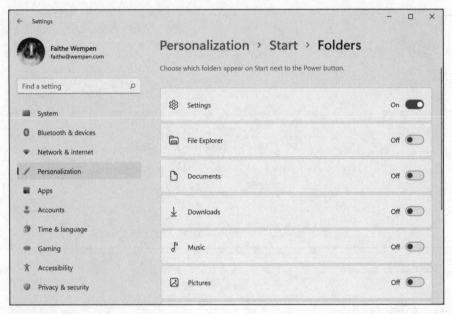

FIGURE 7-2

Customize the Windows 11 Taskbar

You can customize the Windows 11 taskbar by adding and removing pinned icons on it, changing which icons appear at the far-right end of the taskbar (also known as the notification area), and lots more.

To customize the taskbar, right-click the taskbar and choose Taskbar settings. This opens the Personalization ⇨ Taskbar section of the Settings app, shown in Figure 7-3.

From here you can scroll through all the available settings and make your selections. There are four groups of settings here:

» **Taskbar items:** You can turn on/off the display of the basic set of items, such as Search, Task view, Widgets, and Chat.

» **Taskbar corner icons:** You can turn on/off the display of some specialty icons that are useful only if you use certain features, like Pen menu, Touch keyboard, and Virtual touchpad. Most people leave all these off.

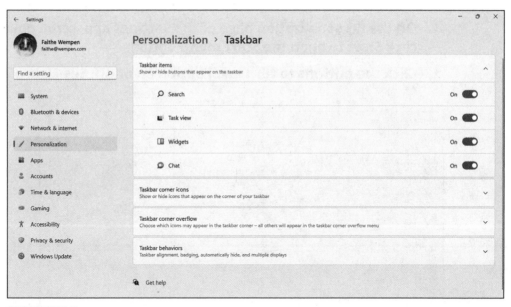

FIGURE 7-3

» **Taskbar corner overflow**: You learned earlier that the notification area shows only a few icons by default, and the others appear on a pop-up menu if you click the up arrow. In this section, you can choose which icons appear initially.

» **Taskbar behavior:** Here you can control how the taskbar in general behaves. For example, you can choose whether the taskbar is centered (default) or left-aligned, and you can automatically hide the taskbar when it's not in use. (If you do that, moving the mouse pointer to the bottom of the screen brings it into view.)

Customize the Windows 10 Start Menu

You can also customize the Windows 10 Start menu in some of the same ways that you can the Windows 11 version. Follow these steps to check that out:

1. **Right-click the desktop and choose Personalize. Then in the navigation bar on the left, click Start.**

2. On the Personalization page of the Settings app, scroll down and click Start to open the Start menu options.

3. Click the buttons to turn each option on or off. See Figure 7-4.

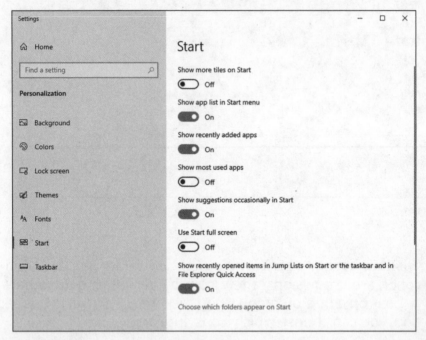

FIGURE 7-4

4. Click Choose which folders appear on Start. A list of folders appears.

This screen enables you to customize which icons appear on the far-left edge of the Start menu. By default, you get Settings, Documents, and Pictures here, but you can add or remove as you like.

5. Close the Settings app window.

You can also customize the Windows 10 Start menu by pinning and unpinning items:

» **Pin:** To pin an app to the Start menu that's not already there, scroll through the alphabetical list of apps, find the one you want, and then right-click it and choose Pin to Start.

» **Unpin:** To unpin an app that's pinned there, right-click the app's pinned icon and choose Unpin from Start.

In Windows 10, you have some additional customization options not present in Windows 11. For one thing, you can change the overall size of the Start menu. Just open it up and then position the mouse pointer on the top or right edge of the menu and drag to change its size. The larger the Start menu is, the more of your pinned tiles you will be able to see at once. Figure 7-5 shows a wider and shorter version of my Start menu. (I don't have much pinned to it.)

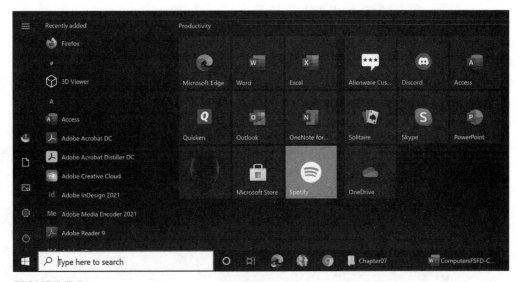

FIGURE 7-5

TIP

You can also resize the pinned tiles on the Start menu. To do that, right-click a tile, point to Resize, and then choose one of the sizes from the menu that appears. You might like to have the tiles for the items you use most often appear larger and the ones you use less frequently be smaller. All the tiles have at least Small and Medium options; some also have Large and Wide options as well.

You can rearrange the tiles by dragging them from place to place. Point to a tile and then press and hold the left mouse button as you move the mouse to drag the tile to another location. To place a tile between two other tiles, drag it between them and then hover there for a moment, still holding the mouse button, and the other tiles will move to make way.

If you drag a tile and drop it directly onto another tile, Windows creates a group containing those two tiles. Groups enable you to cluster tiles together so that the group takes up only one spot. Then when you click the group, it expands.

If you accidentally create a group, that's fixable. Just open the group and then drag each item out of it, one-by-one, until there are no items left. The group disappears automatically when it is empty.

Figure 7-6 shows a customized Start menu with different-sized tiles and a group.

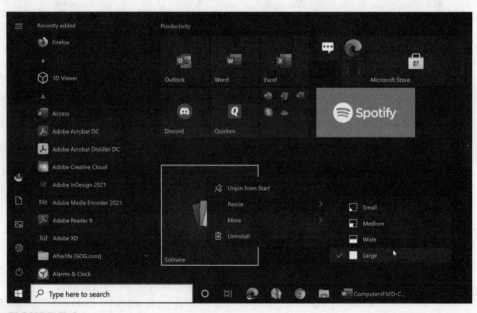

FIGURE 7-6

Customize the Windows 10 Taskbar

As in Windows 11, you can customize the Windows 10 taskbar by adding and removing pinned icons on it, changing which icons appear in the notification area. There are some additional customizations you can make too that Windows 11 doesn't support. (Sensing a pattern here?)

To customize the Windows 10 taskbar, right-click the taskbar and choose Taskbar Settings. This opens the Personalization ⇨ Taskbar section of the Settings app. From here you can scroll through all the available settings and make your selections. There are four groups of settings here:

» **Taskbar:** You can change taskbar behavior settings here, such as automatically hiding it when not in use, using large or small taskbar buttons, and locking the taskbar.

Locking the taskbar means preventing it from being moved to one of the other sides of the screen (top, left, or right). This lock is here because sometimes people accidentally drag the taskbar to move it. Windows 11 doesn't allow the taskbar to be moved so there's no need to have a lock option.

» **Notification area:** This section contains hyperlinks that take you to other screens where you can do two types of customization: Select which icons appear on the taskbar and Turn system icons on or off.

» **Multiple displays:** If you have more than one monitor, the controls in this section adjust how the taskbar appears on monitors other than your main one.

» **People:** Here you can choose to show contacts on the taskbar (most people don't), for easy-click access to contacting certain people.

Customize the Screen Resolution and Scale

Windows can be configured to run in a variety of display resolutions. A **display resolution** is the number of **pixels** (colored dots) vertically and horizontally that make up the display's image. For the sharpest looking image, you want the monitor's highest resolution on most cases. Anything other than the highest resolution available can result in a fuzzy image.

There's a complication with that, though. The higher the resolution, the smaller everything looks onscreen (and the harder it can be to see). So along with the resolution, you will also want to set the **scale**, which is the magnification level of text, icons, and other onscreen items. You can experiment with different scale settings to find the one most comfortable for your vision.

To set the resolution and scale in Windows 11:

1. **Right-click the desktop and choose Display Settings.**

   The Display section of the Settings app appears, as in Figure 7-7.

2. **Open the Display Resolution drop-down list and choose the desired resolution. (The highest setting is usually best. It is probably already set that way.)**

   If you make a resolution change, a box will prompt you to Keep Changes or Revert. If you don't click Keep Changes within 15 seconds, the display reverts to the previous setting. This is done as a precaution in case the new resolution makes your screen unreadable.

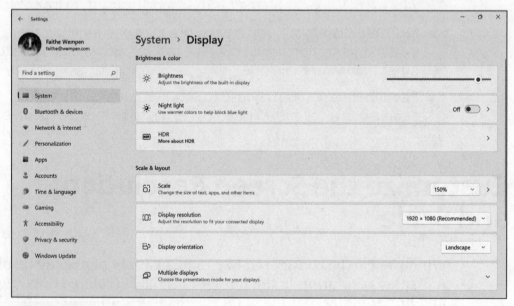

FIGURE 7-7

3. **Open the Scale drop-down list (Windows 11) or the Change the Size of Text, Apps, and Other Items drop-down list (Windows 10) and choose a scale.**

   Higher numbers mean larger stuff onscreen. One setting will be marked (Recommended); that recommendation is for people with normal eyesight. Try the different settings to see what you like best.

4. **Close the Settings app.**

Apply a Desktop Theme

A **theme** is a combination of appearance settings, including a background image, color scheme, system sounds, and mouse cursor appearance. When you apply a theme, you change all those settings at once. (You can also manually change each of them separately.)

Even if you don't like any of Windows' prefab themes, you might still like to know about the Themes feature because you can create your own theme. You can get things just the way you want them, including your own background image and color choices, and then save your settings as a theme. Then you can quickly reapply that theme any time you like. You might have two different themes that you switch between, just to keep things fresh.

To choose a theme, follow these steps:

1. **Right-click the desktop and choose Personalize.**

2. **Click Themes.**

   Thumbnail images for the defined themes appear. The Windows theme is the default, and you can go back to it at any time.

3. **Click a theme thumbnail that appeals to you, and it is automatically applied.**

TIP

To create your own theme, make manual changes to any of the four settings that comprise a theme: background, colors, sounds, or cursor. Then return to the Themes section of the Settings app (repeat steps 1–2) and click the Save button (or Save Theme button in Windows 10). Type a name for your new theme and click Save.

Change Desktop Background Image

Here's where most people really enjoy making Windows their own visually. You can change the background picture and color scheme. For example, my sister has her desktop image set up to be a picture of her grandkids, with a color scheme that matches the logo color of her alma mater. Every time she looks at her desktop, it makes her smile.

To change the background picture on the desktop, follow these steps:

1. **Right-click the desktop and choose Personalize.**

2. **(Windows 11 only) Click Background to display the background options.**

3. **Open the Personalize Your Background drop-down list (or Background in Windows 10) and choose Picture.**

4. **Click Browse Photos (or Browse in Windows 10).**

   The Open dialog box appears, showing the content of your Pictures folder.

5. **Select the picture you want and then click Choose Picture.**

   A sample of that picture appears at the top of the Settings app window. See Figure 7-8.

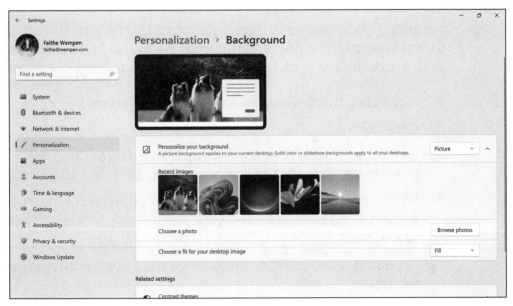

FIGURE 7-8

6. **Open the Choose a Fit for Your Desktop Image drop-down list (or Choose a Fit in Windows 10) and choose a fit. Check out the results in the sample and try different fits until you find the one you like.**

Now that you've seen the basics, here are some more options:

- Choose Solid Color instead of Picture in step 3, and then choose the color you want. When my eyes are tired, sometimes I set the background to a peaceful pastel color like pale green to give myself a break.

- Choose Slide Show instead of Picture in step 3, and then choose an entire folder of pictures that Windows should cycle through at a certain interval. (The default is every 30 minutes.) This can be a great way to not play favorites with pictures of grandkids!

Change the Accent Color

By default, Windows chooses an accent color for you automatically by selecting a color from the background image. The accent color applies to things like the Start menu, taskbar, title bars, and window borders. If you're one of those people who likes to fine-tune every-thing, you might enjoy picking your own accent color. While you're at it, you can also turn transparency on or off on those elements:

1. **Right-click the desktop and choose Personalize.**

2. **Click Colors.**

3. **Open the Choose Your Mode drop-down list (or Choose Your Color in Windows 10) and choose Light, Dark, or Custom.**

4. **If you choose Custom, separate settings appear for your default Windows mode and your default App mode; make your selections.**

5. **(Optional) Click the slider to change the Transparency Effects setting.**

   Transparency makes window title bars and some other elements semi-transparent. This is off by default.

6. **Do one of the following to opt for a manual color choice if needed:**

   Windows 11: In the Accent Color section, open the drop-down list and choose Manual.

   Windows 10: In the Choose Your Accent Color section, clear the Automatically Pick an Accent Color from My Background check box.

7. **Click a color swatch representing the color you want. For more color choices, click View Colors (Windows 11) or Custom color (Windows 10). See Figure 7-9.**

8. **(Optional) To change which elements your color choice applies to, scroll down (if needed) and then select or deselect the options as follows:**

   Windows 11:

   - Set Show Accent Color on Start and Taskbar.

   - Show Accent Color on Title Bars and Window Borders.

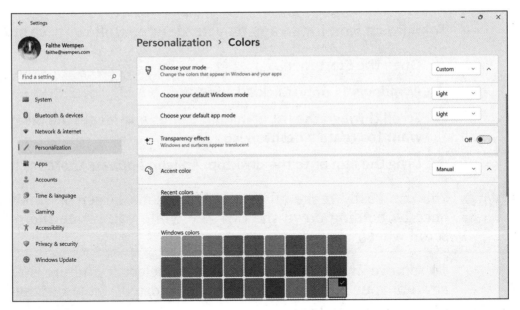

FIGURE 7-9

Windows 10:

- Start, Taskbar, and Action Center.
- Title Bars and Window Borders.

9. **Close the Settings app (or minimize it) and admire your new color choice.**

Manage Desktop Icons

The desktop layout starts out pretty simple, with a couple of icons on it: Recycle Bin and Edge. (There might be others, depending on how the company that made your PC set things up.) You can pump things up by placing your own shortcut icons on the desktop and by rearranging the icons there.

To rearrange icons on the desktop, just drag them where you want them. To line up all your desktop icons in an orderly way, right-click the desktop, choose View, and click Auto Arrange Icons. Note that doing so doesn't just align them initially — it keeps them aligned until you repeat the command to toggle the feature off again.

To place an icon for an app on your desktop, follow these steps:

1. **Open the Start menu.**
2. **(Windows 11 only) Click All apps.**
3. **Scroll through the list of installed apps and locate the one you want to create a desktop icon for.**
4. **Drag the app onto the desktop. An icon appears there for it.**

TIP

You can easily make all the desktop icons larger or smaller at once by holding down the Ctrl key while you rotate the mouse scroll wheel.

To remove an icon from the desktop, select it and press the Delete key on your keyboard. Deleting a shortcut icon on the desktop doesn't delete the app.

WARNING

It is also possible to store actual files on the desktop (not just shortcuts to files). If you delete an actual file from the desktop, it is gone. (Well, it's gone to the Recycle Bin; you could retrieve it from there if you made a mistake.)

Add Widgets to the Desktop

Here's a Windows 11–only feature that you might find handy. **Widgets** are Internet-enabled mini-applications that retrieve specific information quickly. If there's something you ardently pay attention to, like the weather or the stock market, you can place a widget that will keep updated information about it constantly available throughout the day.

Follow these steps to get started with widgets:

1. **Click the Widgets icon on the taskbar.**

   The Widgets panel appears. Any widgets you already have appear at the top.

If the Widgets icon doesn't appear on the taskbar, open the Start menu and start typing **Widgets** and then select it from the search results.

2. **Click Add Widgets.**

   The Widget Settings panel appears.

3. **Click the plus sign next to the widget you want to add. Repeat for other widgets as desired.**

4. **Close the Widget settings panel by clicking its Close (X) button.**

   The main Widgets panel is still open. Each of your widgets shows the latest information about its topic. See Figure 7-10.

5. **(Optional) To get more detailed information about a widget's content, click the widget to open a web page about it.**

FIGURE 7-10

Make Windows More Accessible

People have a variety of physical challenges, such as limited vision, hearing, or mobility. The challenges are as unique as the people — somebody might only be able to use a mouse, for example, whereas another person might only be able to use a keyboard. Fortunately, Windows lets you customize your accessibility features to pick and choose what you need.

If you have accessibility challenges, your best bet is to browse through the available options in the Settings app to see if there is anything you can benefit from. In Windows 11, the section is called Accessibility; in Windows 10 it's called Ease of Access. After you get into that section of the Settings app, explore. Trying out the different features is the best way to learn about them. Figure 7-11 shows the Windows 11 version.

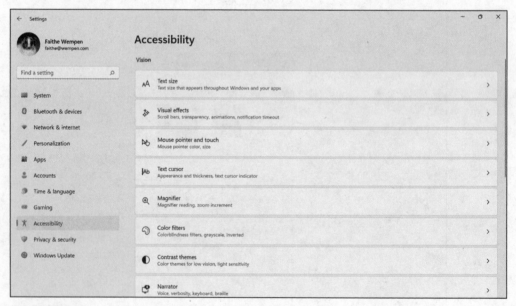

FIGURE 7-11

Table 7-1 lists just a few of the most popular ones and what they can do. The Instructions column's directions assume you are at the top-level Accessibility or Ease of Access screen in the Settings app.

TABLE 7-1 **Windows Accessibility Features**

Feature	Description	Instructions	Keyboard Shortcut
Narrator	Reads onscreen text aloud and announces the actions you take.	Click Narrator. Set Narrator slider to On. Adjust narrator voice in Narrator's voice section.	Windows key + Ctrl + Enter.
Magnifier	Magnifies parts of the screen for easier reading.	Click Magnifier. Set Magnifier slider to On. Set the zoom level by clicking the plus and minus buttons in the Change zoom left section. Adjust magnifier settings as desired.	Windows key + " to turn on Magnifier. Windows key + Ctrl + M to open the Magnifier settings.
Speech Recognition (microphone required)	Enables you to dictate text into applications and give voice commands to control Windows.	Click Speech. Windows 11: Set the Windows Speech Recognition slider to On. Windows 10: Set the Turn on Speech Recognition slider to On. In the Set up Speech Recognition dialog box that appears, follow the prompts.	Windows key + Ctrl + S. To enable voice typing at any time, press Windows key + H or select the Mic button on the onscreen keyboard.

(continued)

TABLE 7-1 *(continued)*

Feature	Description	Instructions	Keyboard Shortcut
Sticky Keys	Enables you to press keys in keystroke combinations one at a time, rather than simultaneously.	Click Keyboard. Set the Use Sticky Keys slider to On.	Press Shift 5 times in a row.
Toggle Keys	Sets up Windows to play a sound when you press Caps Lock, Num Lock, or Scroll Lock so you will notice if you press them by mistake.	Click Keyboard. Set the Use Toggle Keys slider to On.	Press and hold Num Lock for 5 seconds.
Filter Keys	Changes the repeat rate to adjust for keys being pressed very lightly or held down so long it activates twice.	Click Keyboard. Set the Use Filter Keys slider to On.	Press and hold right Shift key for 8 seconds.
On-Screen Keyboard	Provides a window with number, letter, and symbol buttons you can click instead of typing on a keyboard.	Click Keyboard. Windows 11: Set the On-screen keyboard slider to On. Windows 10: Set the Use the On-Screen Keyboard slider to On.	Windows key + Ctrl + O.

Feature	Description	Instructions	Keyboard Shortcut
High Contrast	Makes text and apps easier to see by using more distinct colors.	Windows 11: Click Contrast Themes. Open the Contrast themes drop-down list and choose a theme. Click Apply. Windows 10: Click High contrast. Set the Turn on high contrast slider to On. Open the Choose a theme drop-down list and choose a theme.	Left Alt + left Shift + Print Screen.
Text Size	Changes the size of all onscreen text to make it easier to see.	Windows 11: Click Text Size and then drag the Text size slider to the desired setting. Windows 10: Click Display and then drag the Make text bigger slider to the desired setting.	
Mouse Keys	Enables you to use the numeric keypad on your keyboard to move the mouse pointer.	Click Mouse. Set the Mouse keys slider to On.	

(continued)

TABLE 7-1 *(continued)*

Feature	Description	Instructions	Keyboard Shortcut
Text Cursor	Changes the color and thickness of the on-screen text cursor.	Click Text cursor. Set the Text cursor indicator (Windows 11) or Turn on text cursor indicator (Windows 10) slider to On. Adjust the settings as desired.	
Eye control	With an appropriate eye tracker device connected, enables typing by looking at certain characters on an onscreen keyboard.	Click Eye control. Set the Turn on eye control slider to On. Follow the prompts to set up eye control.	

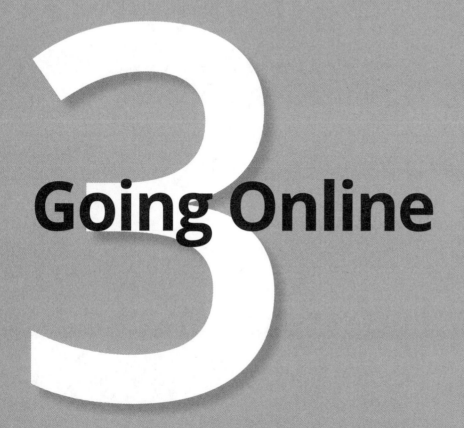

Going Online

IN THIS PART . . .

Getting connected to the Internet

Browsing the web

Staying safe while online

Sending and receiving email

Working with cloud applications

Connecting with people online

Chapter **8**

Getting Connected to the Internet

For many people, going online might be the major reason to buy a computer. You can use the Internet to check stock quotes, play interactive games with others, shop, and file your taxes, for example. For seniors especially, the Internet can provide a way to stay connected with family and friends located around the country or on the other side of the world via email, instant messaging, or video calling. You can share photos of your grandchildren or connect with others who share your hobbies or interests.

But before you begin all those wonderful activities, it helps to understand some basics about the Internet and how it works.

This chapter helps you understand what the Internet and World Wide Web are, as well as some basics about connecting to the Internet and navigating it. You find out what you can do online, and learn how to take an inventory of your computer to make sure you have the hardware and software you need to participate fully in all the online experiences you want to have.

What Is the Internet?

Simply put, the **Internet** is a worldwide computer network that enables different individual networks to communicate with each other. It started as a military communication system, and gradually expanded and became more accessible to private citizens and businesses.

Having a computer lets you do fun things by yourself at home, but the Internet lets you extend that fun into the outside world, venturing out into not just your local neighborhood, but into every corner of the globe.

The Internet is a vast source of all kinds of information, everything from the movie times at your local cinema to the history of Navajo pottery. It's the ultimate public library, except instead of phoning someone at the reference desk, you type in a few keywords and click a link. Billions of websites all over the world serve up all manner of facts, opinions, and sales pitches. In Chapter 9, you learn how to use a web browser, which is your entry point to the information directories and repositories online.

WARNING

Here's an important caveat, though: *not everything you read online is true or objective.* The Internet has democratized information delivery, in that you no longer must be an expert to have your opinion broadcast to millions of people. You must develop critical thinking skills to distinguish uninformed and biased opinions from actual facts you find online. Chapter 10 explains this and other online risks and helps you stay safe and well-informed.

Another important service the Internet offers is communication. There are multiple online ways of communicating with friends, loved ones, businesses, high school classmates — pretty much anyone. The most popular method is electronic mail (email), which is covered in Chapter 11. Chapter 13 looks at some of the other communication methods, like social media, instant messaging, video chatting, and blogging.

You can also run applications that are stored online, and store your important files securely in online storage systems. Both topics fall under the heading of *cloud*, and you learn all about them in Chapter 12.

People and the media bounce around many online-related terms these days, and folks sometimes use them incorrectly. Your first step in getting familiar with the Internet is to understand what some of these terms mean.

Residing on that network of computers is a huge set of documents and services, which form the **World Wide Web**, usually referred to as simply the **web**. It's called a web because there are many interconnected pathways between the servers, so that if one route to a location isn't available, the request takes an alternate route.

The web includes **websites,** such as shopping websites and government websites. These websites are made up of collections of **web pages** just as a book is made up of individual pages. Websites have many purposes: For example, a website can be informational, function as a retail store, or host social networking communities where people can exchange ideas and thoughts.

You can buy, sell, or bid for a wide variety of items in an entire online marketplace referred to as the world of **e-commerce**. For example, you can buy products online and have them shipped to your home, and you can buy software and download it to your computer.

To get around on the web, you use a software program called a **browser**. Many browsers are available, and they're all free. Windows comes with two browsers: Microsoft Edge and Internet Explorer. Browsers offer tools to help you navigate from website to website and from one web page to another, as Chapter 10 explains.

Each web page has a unique address that you use to reach it, called a **uniform resource locator (URL)**. You enter a URL in a browser's address bar to go to a website or to a particular page within a site. An example of a URL is www.dummies.com.

When you open a website, you might see colored text or graphics that represent **hyperlinks**, also referred to as **links**. You can click links to move from place to place within a web page, within a website, or between websites. A hyperlink can be a graphic (such as a photo, company logo, or button) or text. A text link is often (but not always) identifiable by colored or underlined text. It may appear underlined all the time, or only when pointed at (as in Figure 8-1).

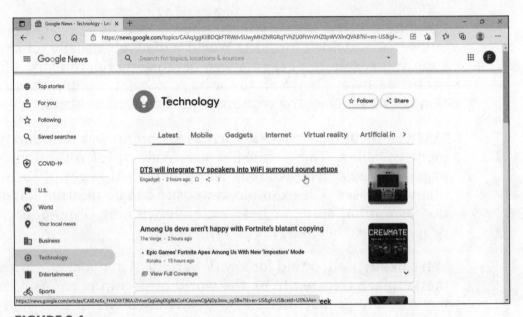

FIGURE 8-1

Explore Different Types of Internet Connections

To get Internet connectivity in your home, you need to sign up with an **Internet service provider** (also referred to as an **ISP** or simply a **provider**). An ISP is a company that owns dedicated computers (called **servers**) that you use to access the Internet. ISPs charge a monthly fee for an Internet connection, and some of them may require an initial commitment of a year or more.

Depending on where you live, you may have a choice of Internet connection types and ISPs that offer them. If you're in a city (or medium-sized town), you will have more choices than someone in a rural or remote area. That's because many of the connection types require a significant infrastructure to be in place, and it's profitable to create and maintain those infrastructures only in areas where there are a lot of people who subscribe.

Often the companies that provide other utilities also offer Internet. For example, a fiber optic or DSL connection might come through your phone or electricity provider, whereas a cable Internet connection might be available through your cable-TV company.

Each connection type offers pros and cons in terms of the quality of the signal and potential service interruptions, depending on the company and your locale, so do your homework before signing on the dotted line. Here are the most common types of connections, listed in rough order of fastest to slowest:

» **Fiber-optic service (FiOS):** This service is delivered through a dedicated fiber-optic cable run directly to your home or office. Phone and electric companies often offer FiOS service, and sometimes fiber-optic television service is also available for an additional cost from the same provider. FiOS is the fastest available Internet service available, but not usually the cheapest.

» **Cable:** If cable TV is available at your home, cable Internet probably is too, from the same company. Cable Internet is relatively fast, with different speed levels usually offered at different price points. Many cable providers offer plans that bundle TV, phone, and Internet services.

» **Digital subscriber line (DSL):** This service is delivered through the same cables as residential phone land lines. Your phone is available to you to make calls even when you're connected to the Internet. There are many different types of DSL, each with its own maximum upload and download speeds, so shop carefully and don't assume that two different DSL providers offer the same level of performance.

» **Satellite:** Especially in rural areas, satellite Internet providers may be your only broadband option. This requires that you install a satellite dish with a special transmitter on it. HughesNet, dishNET (from Dish Network), and Exede (from WildBlue) are three providers of satellite connections to check into. Satellite has some drawbacks, such as **latency** (delay) in data transfers that can hamper the performance of anything you do online that requires lightning-fast response, like playing an online shoot-em-up game. Most satellite providers also cap the amount of data you can transfer per month before slowing down your service or charging you extra.

» **Cellular:** If you have a smartphone that includes Internet service with your monthly plan, you can connect your phone to your computer to piggyback off the phone's Internet connectivity. You can also buy your own portable Wi-Fi hotspot from a provider, so that you can use the hotspot rather than your own phone. This type of connection also requires a data plan with the mobile carrier. Cellular Internet service is often not as fast as other types (although if you have 5G service, it comes pretty close). You still may have issues with latency, though, like with satellite.

» **Wi-Fi hotspots:** If you take a Wi-Fi–enabled laptop computer, tablet, or smartphone with you on a trip, you can piggyback on somebody else's connection. You will find Wi-Fi hotspots in many public places, such as airports, cafes, and hotels. In some cases, the hotspot will be free for anyone to use; in other cases, you must show that you are entitled to use it (for example, at a hotel you might have to provide your name and room number). At a hotel, you may be given a password at check-in that you will need to enter to connect.

Satellite and mobile connections often restrict how much data (data includes websites you view and email you open, for example) you can download per month. If you exceed your quota, your service may slow down dramatically, or you may be charged more. Be sure to monitor your data usage once you get started with a new plan.

» **Dialup:** This is the slowest connection method, and the absolute last resort when nothing else is available. With a dialup connection, you use a phone line to connect to the Internet, entering some phone numbers (local access numbers) that your ISP provides. Using these local access numbers, you won't incur long-distance charges for your connection. However, with this type of connection, you can't use a phone line for phone calls while you're connected to the Internet. And it's *really* slow.

WIRELESS VS. WI-FI

Wi-Fi is one type of wireless networking technology. It refers to the specific type that's designed to transfer data on a home or business computer network. Wi-Fi is sometimes referred to by the standard that specifies its details: IEEE 802.11. There have been various versions of 802.11 over the years; the latest one is 802.11ax.

For the purposes of this chapter the terms wireless and Wi-Fi are synonymous. However, bear in mind that there are other types of wireless networking too, designed for specific purposes and with their own standard numbers. Examples include Bluetooth, which is used to connect wireless peripherals like headphones to computers, and near-field communication (NFC), which is used to read credit cards at sales terminals.

HOW FAST DO I NEED IT TO BE?

The word **broadband** technically refers to any Internet connection that runs at a speed of 1 Mbps (megabits per second) or more. That's still really slow, though, so the term *broadband* has lost most of its original meaning. Today it refers to everything except dial-up.

Broadband Internet connections have different speeds that depend partially on your computer's capabilities and partially on the connection you get from your provider. Before you choose a provider, it's important to understand how faster connection speeds can benefit you.

Faster speeds make everything you do on the Internet happen more quickly. For example, there is less wait when you download a photo to your computer, load a web page, or play an online video. Videos may also play more smoothly, without annoying delays, and video chat systems won't bog down and start distorting the audio or video.

(continued)

(continued)

Broadband services typically offer different plans that provide different access speeds. These plans can give you savings if you're economy minded and don't mind the lower speeds, or offer you much better speeds if you're willing to pay for them. For example, your cable company might have an entry-level package at 25 Mbps, a mid-range one at 76 Mbps, and a top-of-the line package at 150 Mbps.

Are those more expensive plans worth it? That depends on how you use the Internet. If you are going to use the Internet to access video services like Netflix and Hulu, you will want the fastest connection you can afford. It will make your videos play much more smoothly and without delays. If you're primarily interested in email and simple websites, the highest speed available is probably wasted on you.

One way to "right-size" your Internet speed is to start out subscribing to the lowest-speed plan your provider offers, and then upgrade it if you aren't satisfied with your speed. Providers will always cheerfully upgrade you to a more expensive plan.

Identify the Hardware Required

Your ISP should provide all the hardware needed to connect at least one computer to their service. At a minimum, they provide a **broad-band modem,** which is a small box, about the size of a hardback novel, with some flashing lights on it. The ISP runs a cable to your home, and installs a jack in your wall if needed. (If you're getting Inter-net via your phone or cable company, you may already have a usable jack.) Then a cable connects the broadband modem to that jack.

For basic, one-computer Internet connection, you then run an Ether-net cable from the broadband modem to an **Ethernet adapter** in your PC. (That's assuming your computer has an Ethernet adapter. If it doesn't, you can have one installed, or buy a USB model that you can just plug in.) See Figure 8-2. Presto, you're connected to the Internet. An Ethernet jack looks like a wide version of a telephone jack. (The technical name for it is an RJ-45 jack.) The drawback to this method is that the computer needs to sit fairly close to the modem — within cable distance, anyway.

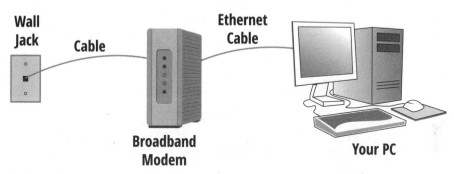

FIGURE 8-2

If you want to make your broadband connection wireless, or share it with multiple devices in your household, you will need a **wireless router**. Instead of connecting one computer directly to the broadband modem, you connect the router to the modem, and then connect each of the computers to the router. (See Figure 8-3.) Your ISP can provide you with a wireless router (for a price), or you can buy one from an office supply or electronics store. A wireless router typically supports both wired and wireless connections, so you have some flexibility on a per-computer basis.

TIP

Some ISPs can supply a combination box that is both a broadband modem and a wireless router. These boxes are nice because they take up less space than two separate boxes would, and ISPs usually rent them to you rather than selling them (which means when you need a newer one, they swap it out for you.)

Each computer that connects to the router must have some type of network adapter. A **network adapter** manages a connection to a network (and the Internet is a network, so it counts). To connect to the router using an Ethernet cable, a computer must have an Ethernet network adapter. To connect to it wirelessly, it must have a wireless network adapter. Most desktop PCs have an Ethernet adapter, and most laptops have a wireless one. You can have additional network adapters installed as needed, or use USB adapters.

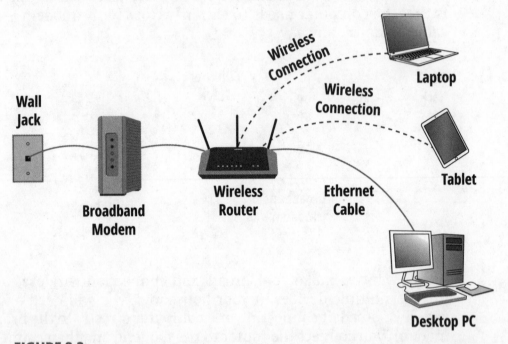

FIGURE 8-3

TIP

If you are buying a new wireless adapter for a computer, look for one that supports the latest Wi-Fi standard (which at this writing is 802.11ax). It's not that other standards won't work, but it costs very little extra to buy yourself some extra time against obsolescence.

If this all sounds like Greek to you, review your computer's system specifications for information about its networking capabilities, and then visit a computer or major office supply store and ask representatives for their advice about your specific hardware.

TIP

Many providers offer free or low-cost setup when you open a new account. If you're not technical by nature, consider taking advantage of this when you sign up.

Set Up a Wi-Fi Internet Connection

Let's suppose you have purchased and installed all the needed hardware, and you're ready to go. You have a wireless router (or a combination box that includes that functionality), you have a computer with a wireless network adapter installed, and now it's time to introduce them to one another.

First, let's check the current status. Look in the notification area (the right end of the taskbar) for the Network icon. It will appear in one of these forms:

 » This symbol means you are already connected to a wireless network. If you hover the mouse pointer over the icon, a ScreenTip tells you which network it is.

 » This symbol means at least one network is available but you aren't connected yet.

If you don't see either of these symbols, you have no network adapters (or at least none that Windows recognizes). You may need to get some help troubleshooting that, either from a tech-savvy friend or a computer repair shop.

TIP

Remember that sometimes not all the icons appear at once in the notification area. Click the up-pointing arrow button in the notification area to see any additional icons before you conclude that you don't see either of them.

But let's be optimistic and assume that you do see the Network icon, and it looks like a globe. Follow these steps to make a connection.

1. **Click the Network icon to open a panel.**

In Windows 11, in the top-left corner is a Wireless icon, with the word Available under it. (See Figure 8-4.)

2. **Click the right-pointing arrow next to the Wireless icon to see a list of the available Wi-Fi networks within range.**

In Windows 10, you immediately see the list of networks; you don't have the extra step of clicking an arrow button.

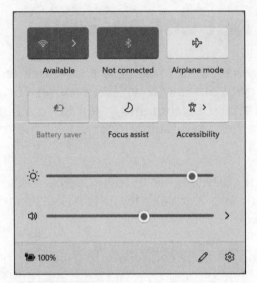

FIGURE 8-4

TIP

If you are near your neighbors, you might see their Wi-Fi networks on this list along with your own. The names on this list are the **service set IDs (SSIDs)** of the wireless routers. If you got your wireless router from your ISP, they may have provided paperwork that tells your router's SSID. If you bought the router yourself, its default name is probably the manufacturer's name. For example, a Linksys router may be named *Linksys*.

3. **Click the SSID of your router to select it.**

A Connect button appears, along with a Connect Automatically check box. See Figure 8-5.

4. **Click to mark the Connect Automatically check box if it is not already marked.**

5. **Click the Connect button.**

6. **If the router has security enabled on it (and it should), and if this is the first time you have connected to it with this PC, a box appears prompting you to enter the network security key. Enter the key and click Next.**

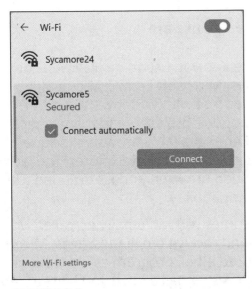

FIGURE 8-5

TIP

If the wireless router came from your ISP, there is probably a sticker on it that lists the security key. If you bought the wireless router yourself, it won't have security enabled unless you enabled it yourself (following instructions that came with the router), in which case you will already know the key.

At this point the list of networks reappears, and beneath the network's SSID you should now see *Connected* or *Connected, secured.*

Assess Your Software Situation

Windows provides all the basic software you need to get started with the Internet, so there's nothing you need to acquire immediately. However, once you get your bearings online, you may choose to upgrade to fancier software that has more features, or get specialized apps that do extra things like video conferencing and online gaming.

The basic set of apps for Internet use are

» **A web browser.** This is the app you use to display web content and access online apps. Microsoft Edge is Microsoft's latest browser, and it comes free with Windows 10 and 11. Google Chrome is a very popular browser, and many people consider it the standard. Other available browsers include Mozilla Firefox, Safari, Opera, and Internet Explorer. The latter is an older Microsoft-provided browser, and it's included with Windows for backward-compatibility.

» **An email app.** Windows comes with an email app called Mail that enables you to send and receive email messages. Microsoft Office includes a full-featured email program called Outlook that you may prefer if you work with email a lot. If you have a web-based email account, like a Gmail or Outlook.com address, you can access your mail via your web browser, so you don't have to open a separate email app.

» **Video-calling app.** If you want to video chat, you will need an app that manages that process. Some common ones include Skype, Zoom, and Google Meet. They're all free to download and install; some of them offer extra features if you subscribe to them. A version of Skype comes with both Windows 10 and 11. Windows 11 includes a version of Microsoft Teams, which includes video chatting among its many features.

Now you're all set to use the Internet! The next several chapters help you get started with the web, email, and other useful and fun Internet activities.

» Searching the web

» Finding content on a web page

» Pinning a tab

» Creating and managing a favorites list

» Using Favorites

» Viewing your browsing history

» Printing a web page

» Creating collections

» Opening an InPrivate browsing session

» Customizing the New Tab page and the Home page

» Adjusting Microsoft Edge settings

Chapter **9**

Browsing the Web

A **browser** is a program that you can use to move from one web page to another, but you can also use it to perform searches for information and images. Most browsers, such as Microsoft Edge, Google Chrome, and Mozilla Firefox, are available for free. Macintosh computers come with a browser called Safari preinstalled with the operating system.

In this chapter, you discover the ins and outs of using Microsoft Edge — the browser included with Windows 10 and 11.

By using Microsoft Edge, you can

» **Navigate all around the web.** Use the Microsoft Edge navigation features to go back to places you've been (via pinned tabs and the Favorites and History features), and use Bing or Cortana to search for new places to visit.

» **Take advantage of reading and note-taking feature.** You can create a list of articles you'd like to read later, view them in Reading view for a more natural reading experience, or make your own notes on a page.

» **Print content from web pages.** When you find what you want online, such as an image or article, just use Microsoft Edge's Print feature to generate a hard copy.

» **Customize your web-browsing experience.** You can modify some Microsoft Edge features such as the default home page to make your online work easier.

Meet the Edge Browser

You can start Edge by clicking the Edge icon pinned to the taskbar. There may also be an Edge shortcut icon on the desktop, and you can also start Edge from the Start menu, like any other application. Figure 9-1 shows the Edge screen.

TIP

Pay no attention to the background graphic that fills most of the Edge screen in Figure 9-1; this is just one of an ever-changing array of starting images that Edge shows. The one you see will be different. After you browse to a web page, or perform a search, that graphic goes away, replaced by the content you are viewing.

The Edge browser is designed to be simple and streamlined. At the top is an **address bar**, where you can enter an address (like www.msn.com) to go directly to that site.

Address bar (doubles as a Search box)

Search the Web bnx

FIGURE 9-1

The address bar doubles as a **search box**; instead of typing an address there, you can type search keywords and then press Enter. Search results appear, and you can click on the closest match to view that site.

On the starting page shown in Figure 9-1, there is also a **Search the web box** on the page itself. You can type your search keywords there instead of in the address bar; the result is the same either way.

To the left of the address bar are navigation tools. These are buttons you can click to take certain actions like moving forward and back, reloading a page, and returning to your home page. To the right of the address bar are icons that open various menus and panes that you will learn about throughout this chapter.

Table 9-1 summarizes the icons you see in Figure 9-1.

TABLE 9-1   **Microsoft Edge Icons**

Icon	Name	Keyboard Shortcut	Purpose
←	Back	Alt+left arrow	Navigates to the previously viewed webpage
→	Forward	Alt+right arrow	After using Back, navigates to the page you were viewing before you clicked Back
↻	Refresh	Ctrl+R	Reloads the current webpage
⌂	Home	Alt+Home key	Returns to the New Tab page (or some other page you specify)
☆⊕	Add this page to Favorites	Ctrl+D	Adds the current webpage to the Favorites list
☆≡	Favorites	Ctrl+Shift+O	Displays the Favorites list
⊕	Collections	Ctrl+Shift+Y	Opens a pane where you can collect and organize content you find online
🔘	Account	None	Opens a menu for signing into and managing user profiles
			This icon will show your user picture if you are logged in and you have a picture; otherwise, it will be a generic icon of a person's head and shoulders
...	Settings and more	Alt+F	Opens a menu from which you can manage Edge settings

There are three ways of displaying web content in a browser:

» **Enter an address:** You can type a web address (also called a uniform resource locator, or URL) directly into the Address bar and press Enter. The page immediately appears.

» **Click a link on another page:** You can click a hyperlink on whatever page you are viewing currently, and that link will take you to a different webpage. This is called **browsing**.

» **Search by keyword:** You can search for web content using keywords. When the research results list appears, you can click a link for a page that appears to contain the content you want. (Sometimes it's hard to tell from a page's description.)

You will learn more about these methods later in this chapter. But first, let's have a look at another useful feature in Edge.

Like most other browsers, Edge uses tabs. In Figure 9-1 there is only one tab showing; its name is New tab. (Not a very catchy name, is it?) Tabs are like little folder tabs in a filing cabinet; they enable you to have more than one webpage open at once, and to switch back and forth between them by clicking the desired tab.

To open a new tab in Microsoft Edge, click the New Tab button (+) to the right of the current far-right tab, or press Ctrl+T. Then navigate to the desired webpage on that tab.

To close a tab, click the Close Tab (X) button on the right end of the tab. You can also press Ctrl+W to close the current tab. Right-clicking a tab opens a menu of management options for it; for example, you can close the tab, close all other tabs *except* this one, pin the tab (that is, make it permanently open each time you start Edge), duplicate the tab, and more. See Figure 9-2.

Right-click a tab to open its menu   New Tab button

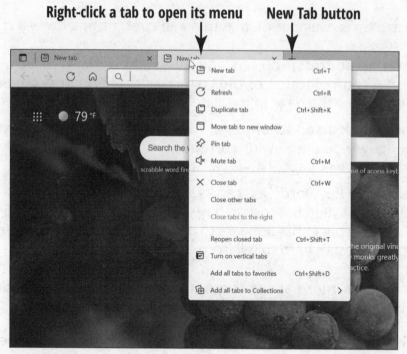

FIGURE 9-2

Search the Web

You can use words and phrases to search for information on the web using a search engine no matter which browser you use. A search engine is a powerful database that you can access using a browser. This database contains information about millions of websites you can visit. When you search for specific keywords, search results appear that list popular sites that match those keywords.

In this example, you'll do two different searches, each one using a different search engine. That way you can see firsthand how the results of a search may vary depending on the engine you choose. You will also practice using the Back, Forward, and Home buttons:

1. Open Microsoft Edge by clicking its button on the taskbar (or using any other method).

2. In the address bar, type a few words that describe what you are looking for. For this example, I will use antique teddy bears. **Then press Enter.**

Search results appear from the Bing search engine, as in Figure 9-3.

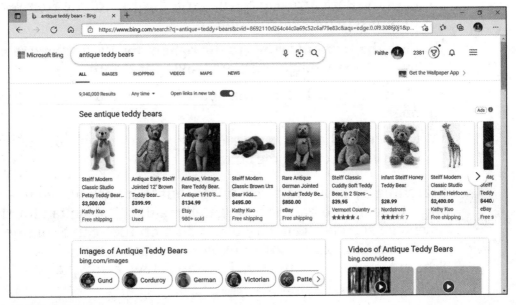

FIGURE 9-3

3. Drag the scroll bar on the right downward to see more of the search results.

4. Click the New tab button (+) above the address bar to open a new tab.

5. Type www.google.com in the address bar and press Enter.

The Google search engine's main page appears.

6. Enter the same keywords that you used in step 2 in the search box in the center of the page, and then press Enter.

Search results appear for the Google search engine.

7. **Click the Bing search's tab to see that list again. Flip back and forth between the two tabs to compare the results.**

   Some of the results will be the same as before; others will be different, or in a different order.

8. **Click one of the links in the search result to visit that page.**

9. **Click the Back arrow in the upper-left corner of the Edge window to return to the search results page.**

10. **Click the Forward arrow (to the right of the Back arrow) to move forward to the page you were looking at before.**

11. **Click the Home button (the house icon) to return to the New Tab page.**

    Later in this chapter, you learn how to customize what page appears when you click the Home button.

TECHNICAL STUFF

Searching by keyword(s) is simple, but there are more advanced ways to search to get more targeted results. The syntax for these advanced searches can vary depending on the search engine you are using. You may want to explore a search engine's Advanced Search features on your own. Here are a few techniques that work with most search engines:

» Putting a plus sign between two keywords will produce results that contain both words. For example, you might search for **teddy+bear**.

» Adding the word *NOT* before a keyword will exclude results that contain that word. For example, you might search for **bear NOT grizzly**.

» Putting a phrase in quotation marks will produce results where those words appear in that exact order with nothing between them. For example, you might search for **"teddy bear picnic"**.

Find Content on a Web Page

In addition to searching for web pages, you can also search for specific content on the active page. For example, if you are viewing a page that contains an entire screenplay or book manuscript, you might want to search for certain words without having to skim the entire page.

To search on a page, follow these steps:

1. **Click Settings and More ⇨ Find on Page, or press Ctrl+F.**

The Search box appears in the upper-right corner of the page.

2. **Type the word or phrase you are seeking into the Search box and press Enter.**

The first instance of it appears highlighted on the page. See Figure 9-4.

3. **Click the down arrow on the Search box to jump to the next instance. Continue doing that until you have seen all the instances (or at least seen all you need to).**

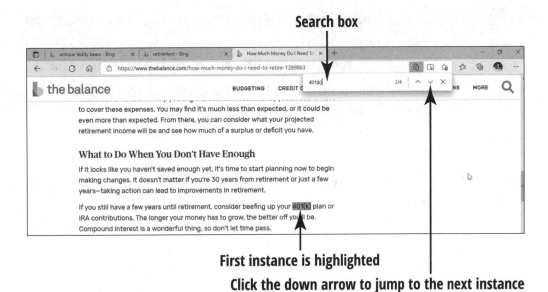

Search box

First instance is highlighted

Click the down arrow to jump to the next instance

FIGURE 9-4

Pin a Tab

If you have a favorite website that you access all the time, one option is to pin a tab so that the tab appears every time you open your browser. When you pin a tab, it appears to the left side of the tabs display with only an icon, no title or Close button. To pin a tab:

1. Display the page you want to pin.

2. Right-click the tab to display its menu (see Figure 9-2) and then click Pin Tab.

To unpin a tab, right-click the tab again and then select Unpin Tab.

Create and Manage a Favorites List

If there's a site you intend to revisit, you may want to save it to Microsoft Edge's Favorites list so you can easily go there again.

To add an item to your Favorites list, follow these steps:

1. Display the page that you want to add to your Favorites list.

2. Click the Add This Page to Favorites button at the right end of the address bar. (It looks like a star with a plus sign on it.)

3. Modify the name of the Favorite listing to something easily recognizable, as shown in Figure 9-5.

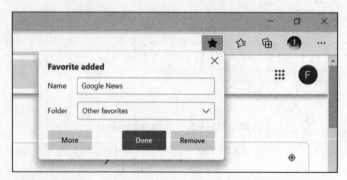

FIGURE 9-5

4. **If you wish, choose another folder from the Folder drop-down list to store the favorite in.**

   The default is Other Favorites. An alternative is Favorites bar, which places it on an optional toolbar that runs across the top of the browser window.

5. **Click the Done button to finish adding the site to the list.**

TIP

The Favorites bar, which I mentioned in step 4, is a handy optional toolbar that appears across the top of the Edge window, providing shortcut buttons for whatever pages you put on it. To toggle it on or off, you can press Ctrl+Shift+B, or you can follow these steps:

1. **Right-click the title bar of the Edge window and choose Customize Toolbar.**

   A Settings tab appears.

2. **On the Settings tab, scroll down to the Customize Toolbar section.**

3. **Next to Show Favorites bar, open the drop-down list and choose Always.**

4. **Close the Settings tab.**

Figure 9-6 shows the Favorites bar enabled. Notice that at the far-right end of the Favorites bar is an Other Favorites button. Clicking it opens the Other Favorites part of the Favorites list. (Remember from Step 4 that Other Favorites is the default save location for items.)

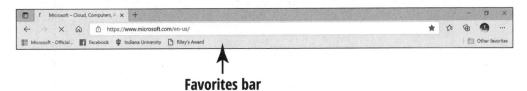

Favorites bar

FIGURE 9-6

To remove an item from the Favorites list, display the page and then click the blue star at the right end of the address bar. (It's official name is Edit Favorite for This Page, and its keyboard shortcut is Ctrl+D.) In the Edit Favorite dialog box that appears, click Remove. See Figure 9-7.

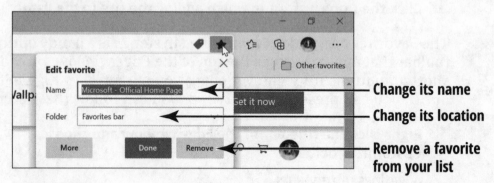

Change its name

Change its location

Remove a favorite
from your list

FIGURE 9-7

You can also edit a favorite from that same dialog box. After opening the dialog box, change the name in the Name box, and/or change the location in the Folder drop-down list.

If you want to create more folders to better organize your favorites, click the More button. This opens a larger version of the Edit Favorite dialog box, shown in Figure 9-8. From here, you can click New Folder to create a new folder. In this same dialog box, you can right-click any existing folder and choose Edit to change its name, or choose Remove to delete it. (Be careful with that, though — any favorites in the folder will be deleted too.)

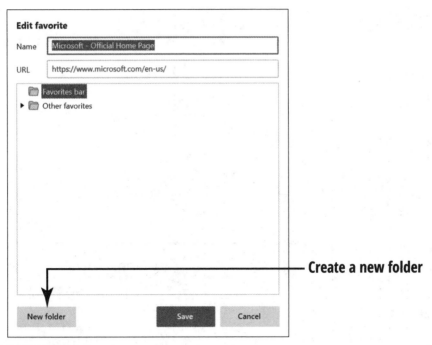

Create a new folder

FIGURE 9-8

Use Favorites

After all that detail about creating favorites, using one is a breeze. Just open the Favorites list and click the page you want to display.

There are two ways to access the Favorites list. You can click the Favorites button in Edge (the star with the three horizontal lines on its right side) to see the entire Favorites menu (both the Favorites bar and Other Favorites), as shown in Figure 9-9.

Alternatively, you can display the Favorites bar (as I explained in the previous section) and then click the Other Favorites button on the bar to see a menu of just the items in Other Favorites. (You don't need to see the Favorites bar items on this list because they're already shown on the bar itself.)

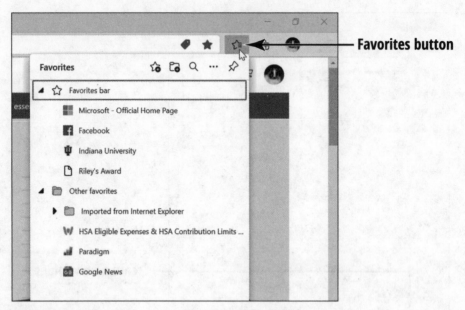

Favorites button

FIGURE 9-9

TIP

Regularly cleaning out your Favorites or reading list is a good idea — after all, do you really need the sites that you used to plan last year's vacation? With the Hub displayed, right-click any item and then choose Delete to remove it.

View Your Browsing History

If you went to a site recently and want to return there but can't remember the name, you might check your browsing history to find it. To view your history, follow these steps:

1. **Click the Settings and More button (. . .) to open its menu and then click History. A shortcut is Ctrl+H.**

   As the History list accumulates items, it groups them by date, as you can see in Figure 9-10.

2. **Click an item to go to it. The History pane closes.**

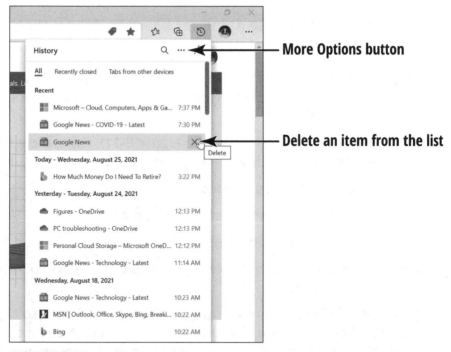

More Options button

Delete an item from the list

FIGURE 9-10

TIP

To delete a single item from the History list, move the mouse pointer over it, and click the Delete (X) that appears at the right. (Refer to Figure 9-10.) You can also right-click the item and click Delete.

To remove all the history items, click the More options button (. . .) near the upper-right corner of the History pane. On the menu that appears, click Clear Browsing Data.

Print a Web Page

If a web page includes a link or button to print or display a print version of a page, click that and follow the instructions. A lot of recipe sites have that feature, for example.

If the page doesn't include a link for printing, follow these steps to print the page:

1. **Click the Settings and More button on the Microsoft Edge address bar, and then click Print in the menu that appears. A shortcut is Ctrl+P.**

2. **In the resulting Print window (see Figure 9-11), choose a printer from the Printer drop-down list, if needed.**

3. **Set any print options as needed:**

   Copies: Type a number in the text box. The default is 1.

   Layout: Click Portrait or Landscape.

   Pages: If you don't want all pages, type the page numbers you want in the text box underneath All.

4. **(Optional) For even more printing options, scroll down to the bottom of the left pane and click More Settings. Then scroll down even further and set the paper size, scale, pages per sheet, margins, and other options.**

5. **Click Print.**

Choose a printer

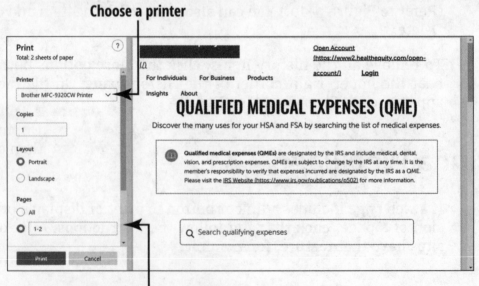

Scroll down for more options

FIGURE 9-11

Create Collections

Collections is a unique feature to Microsoft Edge; other browsers don't have it. It enables you to create a scrapbook-like collection of content from various websites that you visit, including text, pictures, and entire pages. It's kind of like Pinterest or Microsoft OneNote (if you happen to use either one of those).

Click the Collections button (which looks like two squares with a plus sign on one of them) to open the Collections pane. The first time you open Collections, you might be prompted to click through a few information screens that help you understand how to use it. But eventually you will arrive at a blank collection pane that says *Add Content Here.*

Click the Pin button (which looks like a pushpin) at the top of the Collections pane, so it stays open. Then navigate to a page that contains content you want to keep.

You can then click Add Current Page to add the entire current page to the collection (via its address), or you can select and then drag-and-drop text or graphics from the current page into the Collections pane. Figure 9-12 shows a collection with a graphic, a block of text, and an entire webpage.

You can also add notes to a collection. Click the Add Note icon at the top of the Collections pane and then type your note.

When you are done working with the Collections pane, close it by clicking its Close button (X) or unpin it by clicking the Unpin button (which looks like a pushpin with a line through it) and then click the Collections button to toggle it off.

Don't you wish you had access to something like this when you were in school? Kids today don't know how good they've got it.

FIGURE 9-12

Customize the New Tab Page and the Home Page

When you open Microsoft Edge, the first page you see by default is the **New Tab page**. This is also the page that appears every time you open a new tab, which you just learned how to do.

The New Tab page has several layout choices, and each one is very different. The default layout is Inspirational, which you saw back in Figure 9-1. Your other choices are

» **Focused:** You get a simple search box in the center of the page.

» **Informational:** You get a news feed.

» **Custom:** You can customize many aspects of the page's layout.

To choose a New Tab layout, click the Page Settings button (it looks like a little cog) in the top-right corner of the Start page. In the menu that appears (see Figure 9-13), you can change the page layout.

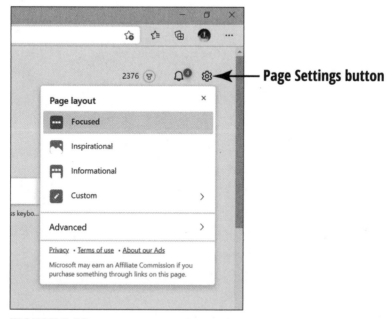

FIGURE 9-13

You can also control whether the New Tab page appears at all when Edge starts; it doesn't have to. Instead, you can reopen the tabs from the previous session, or you can open specific other pages. Follow these steps to check out those options:

1. Click the Settings and More button (. . .) in the top-right corner of the Edge window.

A menu opens.

2. Click Settings.

A Settings tab opens.

3. In the navigation bar on the left, click Start, Home, and New tabs.

4. **In the When Edge Starts section, choose what you want to happen.**

The default is Open the New Tab Page. The alternatives are to Open Tabs from the Previous Session or Open These Pages. See Figure 9-14.

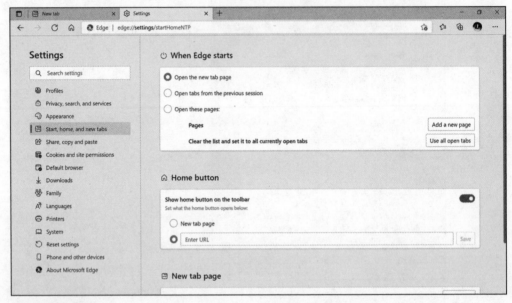

FIGURE 9-14

5. **If you chose Open These Pages, click Add a New Page, and then enter the address of the desired page and click Add.**

The page's name appears below the Pages heading. You can add as many pages as you like here; they will all open automatically when Edge opens.

6. **When you are finished customizing the settings, close the Settings tab by clicking its Close tab (X) button.**

You can also control what happens when you click the Home button (the little house icon to the left of the address bar). By default, it displays the New Tab page, but you can set it to display some other page instead. In Figure 9-14, check out the Home button section, where you can choose New Tab page or type a specific URL (address).

Adjust Microsoft Edge Settings

You can adjust how Microsoft Edge works in many different ways. Click the Settings and More button at the far-right end of the Microsoft Edge address bar, and then click Settings in the menu that appears. Then just start exploring the various settings you find there. There are dozens of them, far too many to cover here, so I'm making this project a self-study. Click a category on the left and then scroll through the various settings on the right. Figure 9-15 shows the Appearance category, where I've selected a different color for the Edge window's title bar and border.

Have fun playing with the settings!

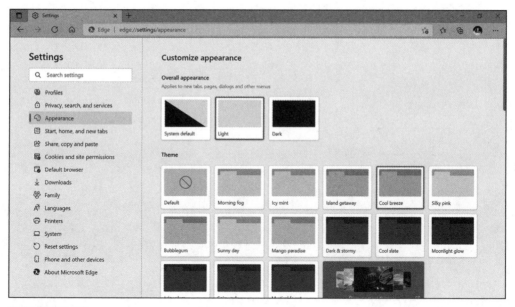

FIGURE 9-15

» Downloading files safely

» Using InPrivate browsing

» Using SmartScreen Filter

» Changing your privacy settings

» Understanding information exposure

» Keeping your information private

» Avoiding phishing scams and other kinds of email fraud

» Creating strong passwords

Chapter **10**

Staying Safe While Online

Getting active online carries with it certain risks, like most things in life. But just as you know how to drive or walk around town safely when you know the rules of the road, you can stay relatively safe online.

In this chapter, you discover some of the risks and safety nets that you can take advantage of to avoid those risks, including

» **Understanding what risks exist:** Some risks are human, in the form of online predators wanting to steal your money or abuse you emotionally; other risks come from technology, such as computer viruses. For the former, you can use the same common sense you use when interacting offline to stay safe. For the latter, there are tools and browser settings to protect you.

» **Being aware of what information you share:** Abuses such as ID theft occur most often when you or somebody you know shares information about you that's nobody's business. Find out how to spot who is exposing information (including you) and what information to keep private, and you'll become much safer online.

» **Avoiding scams and undesirable content:** Use various privacy settings in Windows to limit exposure to undesirable information, such as blocking pop-ups. Also, find out how to spot various email scams and fraud so you don't become a victim.

» **Creating safe passwords:** Passwords don't have to be hard to remember, just hard to guess. I provide some guidance in this chapter about creating passwords that are hard to crack.

Understand Technology Risks on the Internet

When you buy a car, it has certain safety features built in. Sometimes after you drive it off the lot, you might find that the manufacturer slipped up and either recalls your car or requests that you go to the dealer's service department for replacement of a faulty part.

Your computer is similar to your car in terms of the need for safety. It comes with an operating system (such as Microsoft Windows) built in, and that operating system has security features. Sometimes that operating system has flaws — or new threats emerge after it's released — and you need to install an update to keep it secure. And as you use your computer, you're exposing it to dangerous conditions and situations that you have to guard against.

Threats to your computer security can come from a file you copy from a flash drive you connect to your computer, but most of the time, the danger is from a program that you download from the Internet. These downloads can happen when you click a link, open an attachment in an email, or download a piece of software without realizing that *malware* (malicious software) is attached to it.

You need to be aware of these three main types of dangerous programs, which are collectively referred to as **malware**:

» A **virus** is a little program that some nasty person thought up to spread around the Internet and infect computers. A virus can do a variety of things but, typically, it attacks your data by deleting files, scrambling data, or making changes to your system settings that cause your computer to grind to a halt.

» **Spyware** consists of programs responsible for tracking what you do with your computer. Some spyware simply helps companies you do business with track your activities so that they can figure out how to sell things to you; other spyware is used for more insidious purposes, such as stealing your passwords.

» **Adware** is the computer equivalent of telemarketing phone calls at dinnertime. After adware is downloaded onto your computer, you'll see annoying pop-up windows trying to sell things to you all day long. Beyond the annoyance, adware can quickly clog up your computer. The computer's performance slows down, and it's hard to get anything done at all.

To protect your information and your computer from these various types of malware, you can do several things:

» **Run an anti-malware utility.** It's critical that your Windows PC is always running an anti-malware program in the background. It inspects every file you open or save, and warns you of any risks. Some anti-malware software also inspects email and web links. Windows comes with a free, built-in anti-malware tool called Windows Defender, and it's enabled by default. Chapter 18 delves into Windows Defender in detail.

There are also third-party utilities from companies such as McAfee, Symantec, or Trend Micro, and the freely download-able AVG Free from www.avg.com. These third-party utilities have extra features, but most people are fine with the basic Windows Defender tool.

WARNING

People come up with new viruses every day, so it's important that you use software that's up to date with the latest virus definitions and that protects your computer from the latest threats. Thankfully, today's anti-malware software is mostly self-updating. If you use Windows Defender, it is automatically updated via Windows Update. Third-party utilities that run in the background usually download and install their own updates silently and automatically.

» **Use Windows Update to keep Windows current with security features and fixes to security problems.** This is effortless in modern versions of Windows because Windows Update runs automatically (and for the most part it can't be disabled), so you have no choice but to keep Windows up-to-date.

» **Use a firewall.** You can also turn on a firewall, which is a feature that stops other people or programs from accessing your computer over an Internet connection without your permission. Windows comes with a firewall (Windows Defender Firewall), and it's enabled by default. You will learn about it in Chapter 18.

» **Use your browser's privacy and security features,** such as the Suggested Content and InPrivate Browsing features in Microsoft Edge. These settings and features are covered later in this chapter.

Download Files Safely

The two tricks to downloading files while staying safe from malware are to only download from sites you trust and to never download file attachments to emails that you aren't completely sure are safe.

WARNING

The most dangerous files to download are executable files that sport an .exe, .vbs, .com, or .bat extension at the end of the file-name. Clicking on these will run a program of some kind, and could therefore pose an active threat.

Here are some steps you can follow to practice downloading a file from a website:

1. **Open a trusted website that contains downloadable files.**

If you don't know of a file to download but want to practice these steps, try `https://www.google.com/chrome/`. From that site, you can download the very popular browser. You can use it instead of Edge if you like.

2. **Click the appropriate link or button to proceed. (This button often is called Download or Download Now, but on some sites, it is called Install.)**

Windows might display a dialog box asking your permission to pro-ceed with the download; click Yes. The download also, in some cases, might open in a separate Microsoft Edge window.

The file downloads to your personal Downloads folder. A small pane appears in the upper-right corner of the Edge window showing your downloads. See Figure 10-1.

TIP

If you accidentally close that Downloads pane, you can get it back by clicking Settings and more and then clicking Downloads. Ctrl+J is a shortcut.

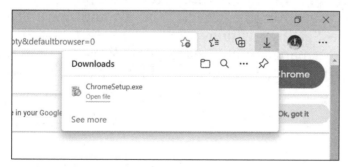

FIGURE 10-1

3. Click Open file.

If you downloaded an executable file, such as a setup file for an application, the setup utility runs. You might see a User Account Control window asking whether you want to allow the app to make changes to your device. Click Yes to continue and then follow the prompts to complete the installation.

If you downloaded some other type of file, such as a picture or video, it opens in the default application for that type of content.

The Downloads folder, which you can open from File Explorer or Edge, is one of your default user folders. Open this folder to view all downloads.

Note that you can download some website picture files by right-clicking and clicking Save Picture. Make sure you have the proper permissions to use any content you download from the web.

If you're worried that a particular file might be unsafe to download (for example, if it's from an unknown source, or if you discover that it's an executable file type, which could contain a virus), cancel the download, or when the User Access Control window asks if the file should be able to make changes to your system, click No.

You may want to explore the Storage Sense feature if you are running low on disk space. It enables Windows to automatically delete files that have been stored in your Downloads folder for more than a certain amount of time (for example, 30 days). It does the same thing for the Recycle Bin. To configure Storage Sense, open the Settings window and select System ⇨ Storage. Under the Storage management heading, click the Storage Sense On/Off setting to On. Then adjust the Storage Sense settings as desired.

Use InPrivate Browsing

Now and then you might want to visit a website that you're suspicious about. Perhaps it doesn't seem quite safe, or you are a little embarrassed about viewing it. For moments like this, Edge provides a feature called InPrivate. When you open an InPrivate browsing window, no history of that session is retained, and the browser security settings are on high alert. The websites you visit during this session will not be allowed to run any apps, save anything to your computer, or make any system changes to your computer.

To start an InPrivate session, click Settings and more (. . .) and then click New InPrivate window. A shortcut is Ctrl+Shift+N. A new Edge window opens, with a dark title bar and border to help you remember that you're working in a special mode. See Figure 10-2.

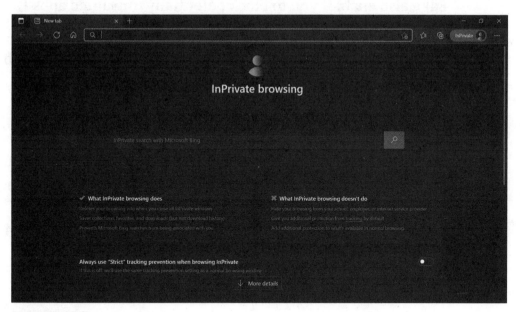

FIGURE 10-2

To close an InPrivate session, just close the browser window. The next time you start Edge, you'll be back to normal.

Use SmartScreen Filter and Block Unwanted Apps

SmartScreen Filter is a security feature in Edge that enables Microsoft to check its database for information on the websites you visit. Microsoft alerts you if any of those websites are known to generate phishing scams or download malware to visitors' computers. The feature is on by default, so you have probably already been using it without even knowing.

TIP

SmartScreen Filter automatically checks websites and will generate a warning message if you visit one that has reported problems. However, that information is updated only periodically, so if you have concerns about a particular site, avoid browsing to it.

Edge also enables you to block potentially unwanted apps from being installed when you visit websites. This feature is on by default too.

Follow these steps to make sure SmartScreen Filter is enabled and potentially unwanted apps are blocked:

1. Click Settings and More (. . .), and then click Settings.

2. Click Privacy, search and services, and then scroll down to the Security section.

3. Check to make sure that the Microsoft Defender SmartScreen option is enabled. See Figure 10-3.

4. Check to make sure that Block Potentially Unwanted Apps is enabled.

5. Close the Settings browser tab.

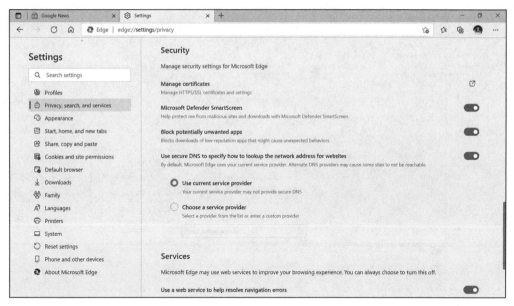

FIGURE 10-3

Change Edge Privacy Settings

You can modify how Microsoft Edge deals with privacy settings to keep information about your browsing habits or identity safer, and to control how cookies are handled. **Cookies** are small text files that specific websites save to your hard drive to retain your settings when you revisit the same site in the future. There are two kinds. **First-party cookies** are used by the website you are visiting, and are usually good. **Third-parties cookies** are used by advertisers, and are usually bad (or at least there is nothing to be gained by accepting them).

In Microsoft Edge, follow these steps to check and potentially change your privacy settings, including cookie settings:

1. Click the Settings and More button, and then click Settings.

2. Click Privacy, Search and Services.

3. In the Tracking Prevention section, choose your preferred level of tracking prevention: Basic, Balanced, or Strict. See Figure 10-4.

Tracking prevention thwarts advertisers from showing you custom-selected apps based on your web activity.

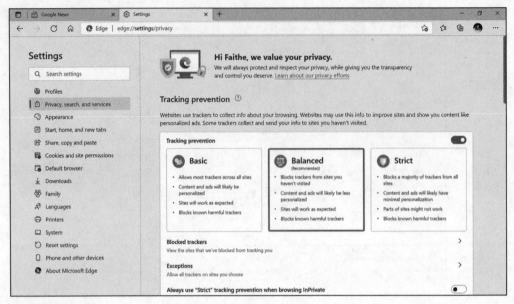

FIGURE 10-4

4. **Scroll down to the Privacy heading, and make sure that Send "Do Not Track" requests is enabled. This alerts any websites you visit that you do not wish to have your activity tracked and sold to advertisers.**

5. **Also under the Privacy heading, enable or disable the Allow Sites to Check If You Have Payment Methods Saved option, according to your preference.**

   When enabled, this setting permits websites that accept credit cards to read your stored information, so you don't have to re-input your credit card information at each site.

6. **In the navigation pane, click Cookies and Site Permissions.**

7. **Click Manage and Delete Cookies and Site Data.**

8. **If desired, enable Block Third-Party Cookies. This gives you more privacy, but some websites may not work properly. You might need to come back here and turn the blockage off later.**

9. **Click See All Cookies and Site Data.**

   A list of all the cookies stored on your system appears. Figure 10-5 shows the ones on my system at the moment; you will have a different list.

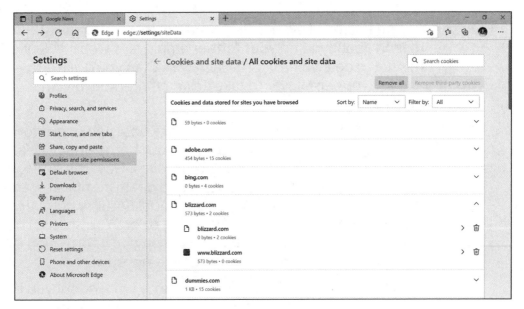

FIGURE 10-5

10. If you want to remove the cookies, click either Remove All or Remove Third-Party Cookies and then click Clear.

If you remove all cookies, you may need to sign back into certain websites that were automatically signing you in before.

WARNING **11.** When you are done adjusting privacy settings, close the Settings browser tab.

Understand Information Exposure

Many people think that if they aren't active online, their information isn't exposed. However, you aren't the only one sharing your information. Consider how others might handle information about you:

» **Employers:** Many employers share information about employees. Consider carefully how much information you're comfortable with sharing through, for instance, an employee bio posted on your company website. How much information should be visible to other employees on your intranet? When you attend a conference, is the attendee list shown in online conference documents? And even if you're retired, there may still be information about you on your former employer's website. Review the site to determine if it reveals more than you'd like it to — and ask your employer to take down or alter the information if needed.

» **Government agencies:** Some agencies post personal information, such as documents concerning your home purchase and property tax on publicly available websites. Government agencies may also post birth, marriage, and death certificates, and these documents may contain your Social Security number, loan number, copies of your signature, and so on. You should check government records carefully to see if private information is posted and, if it is, demand that it be removed.

» **Family members and friends:** They may write about you in their blogs, post photos of you, or mention you on special-interest sites such as those focused on genealogy.

» **Clubs and organizations:** Organizations with which you volunteer, the church you attend, and professional associations you belong to may reveal facts such as your address, age, income bracket, and how much money you've donated.

» **Newspapers:** If you've been featured in a newspaper article, you may be surprised to find the story, along with a picture of you or information about your work, activities, or family, by doing a simple online search. If you're interviewed, ask for the chance to review the information that the newspaper will include, and be sure that you're comfortable with exposing that information.

» **Online directories:** Services such as www.whitepages.com, shown in Figure 10-6, or www.anywho.com, list your home phone number and address, unless you specifically request that these be removed. You may be charged a small fee associated with removing your information — a so-called privacy tax — but you may find the cost worthwhile.

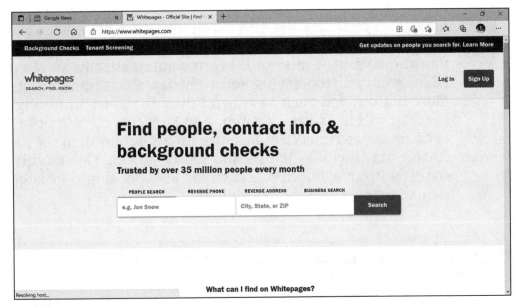

FIGURE 10-6

Online directories often include the names of members of your family, your email address, the value of your home, your neighbors' names and the values of their homes, an online mapping tool to provide a view of your home, driving directions to your home, and your age. The record may also include previous addresses, schools you've attended, and links for people to run background checks on you. (Background check services generally charge a fee.) A smart con artist can use all that information to convince you that he's a friend of a friend or even a relative in distress who needs money.

WARNING

Because services get new information from many sources, you'll need to check back periodically to see if your information has again been put online — if it has, contact the company or go through its removal process again.

TIP

Try entering your home phone number in any browser's address line; chances are you'll get an online directory listing with your address and phone number (although this doesn't work for cellphone numbers).

TIP

Many web browsers not only track browsing data, but also save the personal data you enter into online forms so that information can be reused later to fill forms automatically. In Microsoft Edge, you can stop saving form entries. Click the Settings and More button, and click Settings. Under the Clear Browsing Data heading within Privacy, Search, and Services, click Choose What to Clear Every Time You Close the Browser, and then set Autofill Form Data (Includes Forms and Cards) to On. This might seem counterintuitive, but in this case you are choosing what should be cleared when you close the browser.

Keep Your Information Private

Sharing personal information with friends and family enriches your relationships and helps you build new ones. The key is to avoid sharing information online with the wrong people and shady companies because, just as in the real world, exposing your personal information is one of your biggest risks.

Criminals come in all flavors, but the savvier ones collect information in a very systematic way. Each piece of information is like a series of brushstrokes that, over time, form a very clear picture of your life. And after criminals collect and organize the information, they never throw it away because they may be able to use it many times over.

Fortunately, information exposure is a risk you have a great deal of control over. Before sharing information, such as your date of birth, make sure you're comfortable with how the recipient will use the information:

» **Address and phone number:** Abuse of this information results in you receiving increased telemarketing calls and junk mail. Although less common, this information may also increase a scammer's ability to steal your identity and make your home a more interesting target for break-ins.

- » **Names of husband/wife, father, and mother (including mother's birth name), siblings, children, and grandchildren:** This information is very interesting to criminals, who can use it to gain your confidence and then scam you, or use it to guess your passwords or secret-question answers, which often include family members' names. This information may also expose your family members to ID theft, fraud, and personal harm.

- » **Information about your car:** Limit access to license plate numbers; VINs (vehicle identification numbers); registration information; make, model, and title number of your car; your insurance carrier's name and coverage limits; loan information; and driver's license number. The key criminal abuse of this information includes car theft (or theft of parts of the car) and insurance fraud. The type of car you drive may also indicate your financial status, and that adds one more piece of information to the pool of data criminals collect about you.

- » **Information about work history:** In the hands of criminals, your work history can be very useful for "authenticating" the fraudster and convincing people and organizations to provide him or her with more about your financial records or identity.

- » **Information about your credit status:** This information can be abused in so many ways that any time you're asked to provide this online, your answer should be "No." Don't fall for the temptation to check your credit scores free through sites that aren't guaranteed as being reputable. Another frequent abuse of credit information is found in free mortgage calculators that ask you to put in all kinds of personal information for them to determine what credit you may qualify for.

TIP

Many people set automatic responders in their email, letting people know when they'll be away from their offices. This is helpful for colleagues, but exercise caution and limit to whom you provide the information. Leaving a message that says, "Gone 11/2–11/12. I'm taking the family to Hawaii for ten days," may make your house a prime target for burglary. And you'll probably never make the connection between the information you exposed and the offline crime.

WARNING

You may need to show your work history, particularly on resumes you post on Internet job or business-networking sites. Be selective about where you post this information, create a separate email account to list on the resume, and tell what kinds of work you've done rather than give specifics about which companies and what dates. Interested, legitimate employers can then contact you privately, and you won't have given away your life history to the world. After you've landed the job, take down your resume. Think of it as risk management — when you need a job, the risk of information exposure is less than the need to get a job.

Spot Phishing Scams and Other Email Fraud

As in the offline world, the Internet has a criminal element. These cybercriminals use Internet tools to commit the same crimes they've always committed, from robbing you to misusing your good name and financial information. Know how to spot the types of scams that occur online, and you'll go a long way toward steering clear of Internet crime.

Before you click a link that comes in a forwarded email message or forward a message to others, ask yourself:

» **Is the information legitimate?** Sites such as www.truthorfiction.com, www.snopes.com, and http://urbanlegendsonline.com (see Figure 10-7) can help you discover if an email is a scam.

» **Does a message ask you to click links in email or instant messages?** If you're unsure whether a message is genuinely from a company or bank that you use, call it, using the number from a past statement or the phone book.

REMEMBER

Don't call a phone number listed in the email; it could be a fake. To visit a company's or bank's website, type the address in yourself if you know it or use your own bookmark rather than clicking a link. If the website is new to you, search for the company online and use that link to visit its site. Don't click the link in an email, or you may land on a site that looks right — but is in reality a good fake.

» **Does the email have a photo or video to download?** If so, exercise caution. If you know the person and he told you he'd send the photo or video, it's probably fine to download, but if the photo or video has been forwarded several times and you don't know the person who sent it originally, be careful. It may deliver a virus or other type of malware to your computer.

FIGURE 10-7

In addition to asking yourself these questions, also remember the following:

» **If you decide to forward (or send) email to a group, always put their email addresses on the Bcc: (or Blind Carbon Copy) line.** This keeps everyone's email safe from fraud and scams.

» **Think *before* you click.** Doing so will save you and others from scams, fraud, hoaxes, and malware.

Create Strong Passwords

A strong password can be one of your best friends in protecting your information in online accounts and sites. Never give your password to others, and change passwords on particularly sensitive accounts, such as bank and investment accounts, regularly.

Table 10-1 outlines five principles for creating strong passwords.

TABLE 10-1 **Principles for Strong Passwords**

Principle	How to Do It
Length	Use at least ten characters.
Strength	Mix it up with upper- and lowercase letters, characters, and numbers.
Obscurity	Use nothing that's associated with you, your family, your company, and so on.
Protection	Don't place paper reminders near your computer.
Change	The more sensitive the information, the more frequently you should change your password.

Look at Table 10-2 for examples of password patterns that are safe but also easy to remember.

TIP

You shouldn't use the same password for multiple websites, so it can be a real challenge to remember all your passwords. A lot of people use password management software, which safely stores your passwords on your local PC and recalls them when you visit the corresponding websites. One example is TrueKey by McAfee, which comes with some McAfee security suites.

TABLE 10-2   Examples of Strong Passwords

Logic	Password
Use a familiar phrase typed with a variation of capitalization and numbers instead of words (text message shorthand).	L8r_L8rNot2day = Later, later, not today 2BorNot2B_ThatIsThe? = To be or not to be, that is the question.
Incorporate shortcut codes or acronyms.	CSThnknAU2day = Can't Stop Thinking About You today 2Hot2Hndle = Too hot to handle
Create a password from an easy-to-remember phrase that describes what you're doing, with key letters replaced by numbers or symbols.	1mlook1ngatyahoo = I'm looking at Yahoo (I replaced the Is with 1s.) MyWork@HomeNeverEnds = My work at home never ends
Spell a word backward with at least one letter representing a character or number.	$lidoffaD = Daffodils (The $ replaces the s.) y1frettuB = Butterfly (The 1 replaces the l.) QWERTY7654321 = This is the six letters from left to right in the top row of your keyboard, plus the numbers from right to left across the top going backward.
Use patterns from your keyboard. (See Figure 10-8.) Make your keyboard a palette and make any shape you want.	1QAZSDRFBHU8 is really just making a W on your keyboard. (Refer to Figure 10-8.)

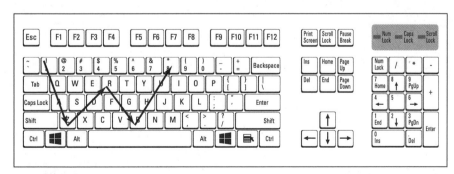

FIGURE 10-8

But you don't have to pay to get help remembering passwords. Microsoft Edge includes a feature that can remember passwords for you. Follow these steps to set it up and check out its settings:

1. **Click the Settings and More button, and then click Settings.**

2. **Click Profiles in the navigation pane.**

3. **Make sure that Offer to Save Passwords is enabled.**

4. **(Optional) Enable the Show Alerts When Passwords are Found in an Online Leak setting.**

   This allows Windows to alert you if any of your passwords saved in Edge are found in a known repository of exposed credentials, so you can immediately change the compromised password.

5. **Scroll down on the page and view your saved passwords. To view a password, click the eye symbol. To remove a password from Edge's list, click the More button (. . .) to its right and click Delete.**

6. When you are finished managing passwords, close the Settings browser tab.

IN THIS CHAPTER

» **Signing up for an email account**

» **Setting up an account in the Mail app**

» **Getting to know the Mail app**

» **Receiving messages**

» **Replying to or forwarding a message**

» **Creating and sending messages**

» **Managing addresses**

» **Sending an attachment**

» **Changing account settings in Mail**

Chapter **11**

Keeping in Touch with Mail

E lectronic mail, or **email** for short, is a system for sending and receiving private messages online. Email is like postal mail, in that you compose it and drop it "in the mail." Then the message is transferred to the recipient's mailbox, and they retrieve it from there and read it.

WARNING

Like postal mail, email is not necessarily an instant delivery system. Your message might travel from your mailbox to theirs in seconds, but then they might not check their mail for several days. Not everyone is glued to their computer 24/7! But what email lacks in immediate gratification, it makes up for in flex-ibility. Email can deliver not only text and hyperlinks to online content, but also attachments. So, for example, you can send an email message with attached word processing documents, pho-tos, spreadsheets, or whatever.

One way to send and receive email is via the web interface associated with your email account. Your ISP will provide a web-based inter-face for any email accounts they supply, and each of the web-based email services also has its own web-based interface. But if you have several email addresses, using their web-based interfaces may mean that you must jump around online, signing into multiple websites to read all your mail. What a pain! Because of that, if you have multiple email accounts, you will probably want to set them up in an email application. Windows comes with one called Mail; that's the one this chapter covers.

This chapter explains the basics: how to get an email address, and how to set it up in Mail. You'll also learn how to send, receive, and forward messages, and how to include attachments and adjust set-tings in Mail.

Sign Up for an Email Account

I'm going to go out on a limb here and say that you already have an email address — because you needed one to sign up for a Micro-soft account, which you need to sign in to Windows. So, hopefully this concept of email is not totally unfamiliar to you. In fact, if you already have an email account that you're happy with, you can skip this section. No need to introduce extra complexity unnecessarily.

Most ISPs provide you with free email accounts for your whole house-hold, to use as long as you subscribe to that ISP's service. Some of them even allow you to keep your email address if you switch to a different ISP (because they are hoping you will come back to their service if you retain that tie to it). Check with your ISP to find out how to access the email account(s) you get for free with their service.

However, because it can be a lot of work to change to a different email address, many people choose to use a free, provider-agnostic email account through a web-based service such as Gmail or Outlook.com. The Mail app in Windows can work with just about any type of email account, so you're free to choose whatever makes the most sense for your situation.

REMEMBER

Email accounts come with certain features. If you are on the fence about which email provider to use, consider these factors:

» **Storage:** Each account includes a certain amount of storage for your saved messages. More is better. Some providers charge extra for anything beyond a minimal amount of storage space. Both Outlook.com and Gmail offer 15 GB with their basic free plan.

» **Address book:** The account should also include an easy-to-use address book feature to save your contacts' information.

» **Personal management extras:** Some email accounts, like Gmail and Outlook.com, offer additional tools beyond email, such as calendars and to-do lists.

» **Junk mail filtering:** Whatever service you use, make sure it has good junk-mail filtering to protect you from unwanted emails. You should be able to modify junk-mail filter settings so that the service places messages from certain senders or with certain content in a junk-mail folder, where you can review the messages and exercise caution about which you open or delete.

» **Anti-malware protection:** Some services include an anti-malware tool that scans incoming messages for potentially harmful attachments and hyperlinks to known-fraudulent websites.

If you don't know what to pick, I suggest Outlook.com, for several reasons. It has all the features listed here, to begin with. It also has the advantage of being hosted by Microsoft, the same company that makes Windows. That means the Mail app is likely to work smoothly and seamlessly with its accounts, and any future updates to Windows will consider Outlook.com compatibility.

Gmail is another good one. It's one of the top web-based email systems in the world. Any future updates to *any* email app will include strong support for Gmail because to do otherwise would alienate a potential userbase of millions of people.

Using your web browser, navigate to `https://www.outlook.com` or `https://www.gmail.com` and follow the prompts there to sign up for your free account.

WARNING

You will have the opportunity to choose your own email address — at least the part to the left of the @ sign. You'll find that most of the "easy" names are already taken, so you might have to get creative with your choice. Whatever you pick, make sure that it doesn't use your full name, your location, age, or other identifiers. Such personal identifiers might help scam artists or predators find out more about you than you want them to know. Also make sure that it's not a moniker that will embarrass you in any context. For example, `sexysenior4you@outlook.com` might be a fun address to have, but when you're corresponding with your grandchildren, you may want something more G-rated. (No, I didn't look to see if that was already taken. It probably is, though.)

Set Up Accounts in the Mail App

Start the Mail app from the Start menu, just like any other app. It's pinned there, so just click the Start button and then click Mail.

TIP

By default, the Mail app attempts to set itself up automatically for the email address you use for your Microsoft account. If you open the Mail app and you immediately see your received messages, that's why. Nice! So it's possible you might not have to do any setup at all (unless you have additional email accounts you need to set up). But if instead you open the Mail app and you see a mostly blank screen, or a Welcome screen, then you have some setup to do.

If a Welcome screen appears, follow the prompts. They should be self-explanatory. I'm being intentionally vague here because Microsoft tends to change the exact steps periodically.

If there's no Welcome screen, follow these steps:

1. **Click Accounts, and then click Add Account.**

The Add an Account dialog box opens. See Figure 11-1.

2. **Do one of the following:**

- If any specific email accounts are suggested at the top of the dialog box, and you see the one you want to set up, click it. These are accounts used elsewhere in Windows, such as with Microsoft accounts used to sign into Windows itself.

- Click the option that best represents the type of email address you have. Scroll down in the dialog box to see more choices.

3. **Follow the prompts that appear. They'll be different depending on what you chose.**

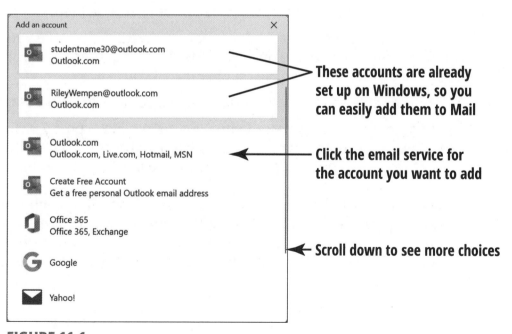

FIGURE 11-1

That's pretty much it! For most email accounts, the process is super easy and doesn't require any tough decisions or complex information to enter.

Let's say, though, that the easy way didn't work for you, maybe because you have an email address that's not from one of the big popular services. Your best bet at this point is to consult the tech support people for your email provider. Tell them you want to set up your account in Mail, and ask them for the information you need.

Get to Know the Mail Interface

The Mail app has a sparse, clean interface, in line with the whole Windows approach. There are three vertical panes, as you can see in Figure 11-2:

» On the left is the Folders pane (also known as the navigation pane), which lists folders and accounts. For example, notice that there are folders for Inbox, Drafts, Send, and Deleted.

» At the bottom of the Folders pane are buttons that take you to other parts of the app, like the calendar, address book, and to-do list. Explore these on your own if you like.

» The center pane is your Inbox, a list of your received messages.

» If no message is selected, the right pane is just a pretty picture. But if one of the messages is selected in the Inbox pane, the message content appears in the right pane.

If you have multiple email accounts set up in Mail, each one has its own separate Folders list. Figure 11-2 shows two accounts, for example. The set of folders you see in the Folders pane depends on which account is selected. When you click an email account in the Folders pane, its folders become visible.

Folders pane   **Inbox**   **Message pane**

Icons for other parts of the app

FIGURE 11-2

TIP

Notice also the generic names that Mail has assigned to each of my two accounts in Figure 11-2: Outlook and Outlook 2. Boring! You can change those names by right-clicking an account and choosing Account Settings. In the Account Settings dialog box, change the name in the Account Name text box and click Save.

Receive Messages

When you open the Mail app, it automatically does a send/receive operation, and any email you've received appears on the Inbox list in the center pane. It automatically repeats this operation every few minutes, but you can also manually trigger it any time by clicking the Sync this View button at the top of the center pane. See Figure 11-3.

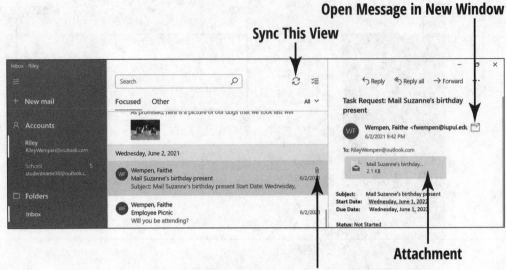

FIGURE 11-3

To view the content of a received message, click the message in the Inbox pane. The message appears in the Message pane on the right.

Notice that the number 5 appears next to the second account in the Folders pane in Figure 11-3. That means there are 5 unread messages. That's handy to know, because when you see a number on an account, you know to click that account to switch over to its folders and see what's new.

To pop out the message into its own window, click the Open Message in a New Window icon, which appears at the top of the Message pane.

A paperclip symbol next to a message in the Inbox indicates that there is an attachment. Some attachment types appear inline with the message, like the message shown in the Message pane in Figure 11-2. (That's true for most common picture formats.) Others appear as just their names, like the message shown in the Message pane in Figure 11-3.

To open an attachment, right-click it in the Message pane and click Open. It opens in whatever application is the default for its file type.

To save an attachment to your local PC, right-click it and click Save. In the Save As dialog box, choose a location and file name and click Save.

Double-clicking an attachment has different effects depending on the attachment type. If the attachment is an email message, the message opens in its own window. If the attachment is any other kind of file, double-clicking is the same as right-clicking and choosing Save.

If you move your mouse over a message in the Inbox, Archive, Delete, and Flag icons appear at the right, as shown in Figure 11-4. Click the Archive icon to save it to an archive file, getting it out of your way without deleting it entirely. Click the Delete icon to delete the message, which moves it to the Trash or Deleted Items folder (which are the same thing; different accounts just name it differently). Click the Flag icon to display a red flag on the message, marking it as a message you want to review later.

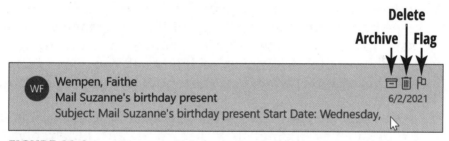

FIGURE 11-4

Reply to or Forward a Message

If you receive an email and want to send a message back, use the Reply feature. Display the message you want to reply to, and then click one of the following reply options at the top of the Message pane, as shown in Figure 11-5:

- **Reply:** Send the reply to only the author.
- **Reply All:** Send a reply to the author as well as to everyone who received the original message.

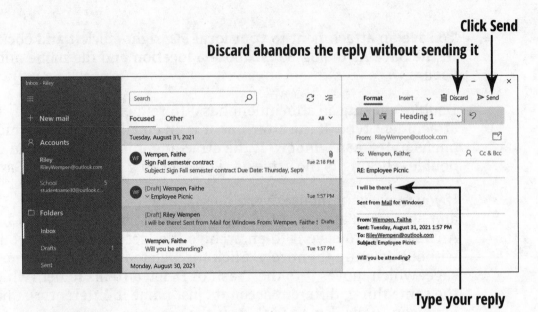

Click Send

Discard abandons the reply without sending it

Type your reply

FIGURE 11-5

When you click either of those options, a reply window opens in the Message pane. (You can pop it out in its own pane by clicking Open Message in New Window.) The recipient(s) are already filled in for you. You can enter additional recipients if you like. (More on that later in this chapter, when you learn how to compose a message.)

Type your reply in the message body, and click Send. Or, if you change your mind about replying, click Discard to cancel the reply.

TIP

If you start creating a message or reply and have it open for a few moments, Mail automatically saves it as a draft. If you click the Back button or close the Mail app without sending the message, Mail adds a copy of the message in the Drafts folder with [Draft] added in the message listing. You can open the message later from there to finish and send it.

To share an email you receive with others, use the Forward feature. Forwarding is like replying except it sends it to a new recipient, not the person who sent the original.

To forward a message, display it, and then click the Forward button at the top of the Message pane. A new message appears in the

Message pane, with the original message quoted. There is no address in the To box; you must fill that in yourself. (I already did that in Figure 11-6.) Then type any message you want to include in the message body and click Send.

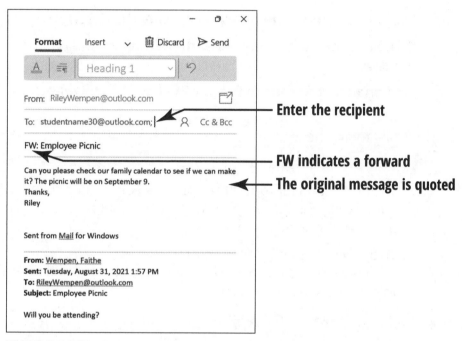

FIGURE 11-6

Create and Send Email

Creating email is as simple as filling out a few fields in a form; all you need is the email address of the person you want to send a message to:

1. In the Mail app, click the New Mail button at the top of the Folder pane.

A new email form opens in the Message pane.

2. In the To box, type the email address of the recipient.

If there are multiple recipients, separate the addresses with semicolons.

TIP

If you want to send a courtesy copy of the message to other people, click the Cc & Bcc link to add those fields to the form, and then enter additional addresses in those boxes. The difference is that with a Cc, all recipients see that person as having been copied, but with a Bcc they don't.

3. **Click in the Subject field and type a concise, descriptive subject.**

4. **Click in the message pane beneath the subject and type your message.**

5. **(Optional) If you want to format the text, do the following:**

   a. At the top of the Message pane, click Format to display the formatting tools if needed.

   b. Drag over the text to select it.

   c. To apply character formatting, click the Font button to open a palette of tools for character formatting (like fonts, sizes, and colors) and make your selection.

   d. To apply paragraph formatting, click the Paragraph button to open a palette of tools for that (with things like bulleted and numbered lists) and make your selection.

   Figure 11-7 shows a message ready to send.

6. **Click the Send button at the top of the Message pane to send the message.**

TIPS FOR COMPOSING EMAIL

Here are some tips for making your email messages the best they can be:

- Don't press Enter at the end of a line when typing a message. Mail and most email programs have an automatic text-wrap feature that does this for you.

- Be concise. If you have lots to say, consider sending a letter by snail mail or overnight delivery. Most people tire of reading text onscreen after a short while.

- Keep email etiquette in mind as you type. For example, don't type in ALL CAPITAL LETTERS. This is considered shouting, which is rude.

- Be polite even if you're really, really angry. Your message could be forwarded to anybody, anywhere, and you don't want to get a reputation as a hothead.

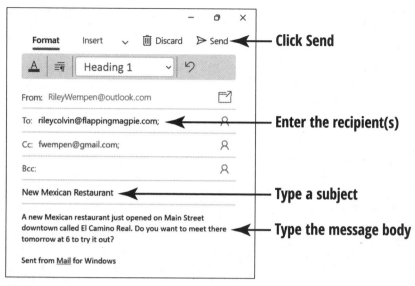

FIGURE 11-7

Manage Addresses

That last example had you manually typing the recipient's email address, but you don't have to memorize everyone's address. That would be nearly impossible, given all the people you want to correspond with. The Mail app includes a very handy address book that is linked to the People section of the Mail app.

As you begin to type an email address in the To, Cc, or Bcc field, Mail cross-references the addresses in the People app and if it finds a match, it suggests it to you. You can click the suggestion to enter it. See Figure 11-8.

You can also click the Choose Contacts button to the right of the To, Cc, or Bcc field to open the People app, shown in Figure 11-9. Then click the person (or persons) you want to send the email to and then click the Done button (the checkmark) at the bottom of the People pane to return to composing the message.

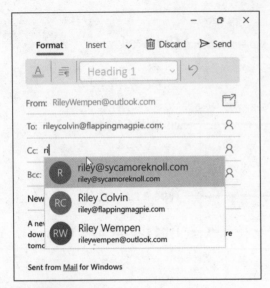

FIGURE 11-8

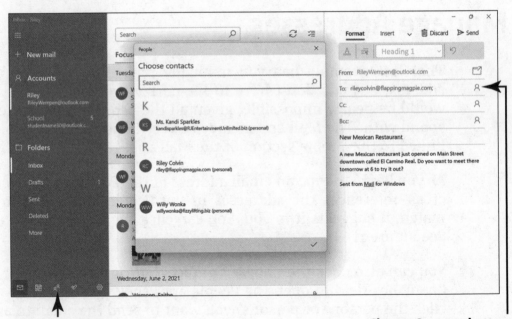

People app button

Choose Contacts button

FIGURE 11-9

TIP

To manage the list of contacts in the People app (for example, to add and delete people and make changes), click the People icon in the lower-left corner of the Mail app's window. See Figure 11-9.

Send an Attachment

It's very convenient to be able to attach a document or image file to an email that the recipient can open and view on his end. To do this:

1. **Start creating the message as usual, with a recipient, subject, and message body.**

2. **At the top of the Message pane, click Insert to display the insertion tools.**

3. **Click Files.**

   The Open dialog box appears.

4. **Locate the file or files that you want and click it (or them). To select more than one file in the same location, hold down the Ctrl key as you click each one. Then click Open to attach them to your message. See Figure 11-10.**

   A thumbnail of the attached file appears above the message body, indicating that it's uploaded.

5. **If you have other attachments from other locations, repeat steps 3–4 as needed.**

6. **Click the Send button to send the message and attachment.**

TIP

You can attach as many files as you like to a single email by repeating steps in this task. Your limitation is size. Various email providers have different limitations on the size of attachments, and some prevent you from attaching certain types of files for security reasons. If you attach several documents and your email fails to send, consider using Microsoft's OneDrive file-sharing service instead. See Chapter 12 for more about using OneDrive.

Click Files   Click Insert

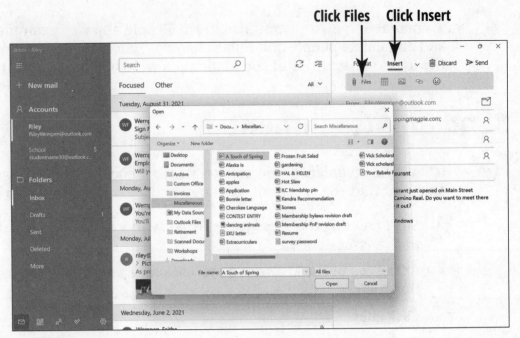

FIGURE 11-10

TIP

If you change your mind about sending an attachment while creating an email, just click the Remove Attachment button (the X) to the right of the attachment.

Change Mail Account Settings

The Mail app has many settings you can adjust for the optimal mail-handling experience. I want to round out this chapter by walking you through the settings that pertain to individual mail accounts:

1. **From within Mail, click the Settings icon at the bottom of the Folders pane, and then click Manage Accounts in the Settings pane (see Figure 11-11).**

2. **Click the account whose settings you want to change.**

   The Account Settings dialog box opens. See Figure 11-12.

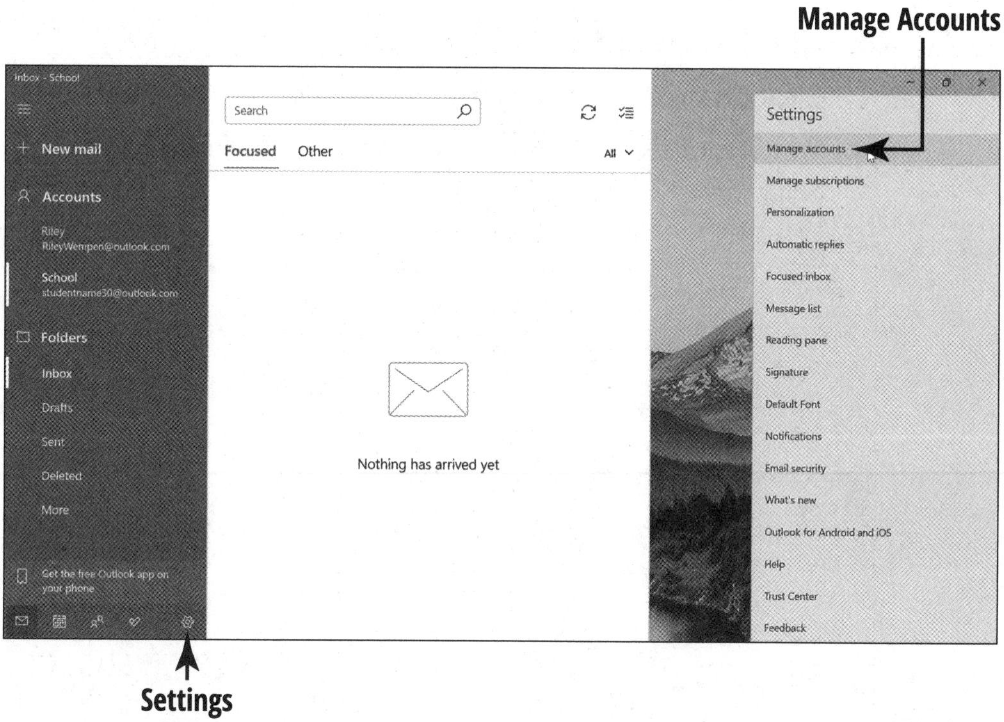

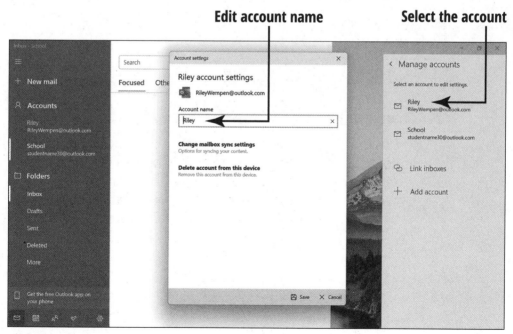

Manage Accounts

Settings

FIGURE 11-11

Edit account name

Select the account

FIGURE 11-12

3. **Edit the Account Name if you want.**

4. **Click Change Mailbox Sync Settings.**

The Account Settings dialog box changes to show the sync settings. See Figure 11-13.

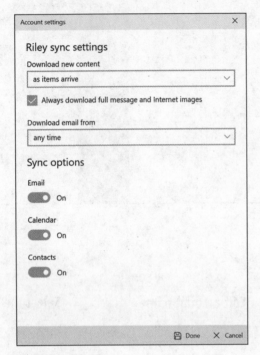

FIGURE 11-13

5. **Change any of the synchronization options as desired:**

- *Download New Content:* You can click this box and, from the drop-down list that appears, choose to download content when a message arrives, every 15 or 30 minutes, or hourly. If you prefer, you can choose to download items only when you click the button beside the account by clicking the Manually option here.

- *Download Email From:* This is a handy setting if you're away from Mail for a while and you've been checking messages in your browser. If so, you may not want to download a month's worth of messages you've already read, so choose another setting from this drop-down list, such as The Last 7 Days.

- *Sync Options:* Set the on/off slider for Email, Calendar, and Contacts as desired. You can synchronize one without the others if you like.

6. **Click the Done button to go back to the Account Settings window (refer to Figure 11-12).**

7. **Click the Save button to finish and close the window.**

To remove an account from Mail, click Delete Account in the Account Settings window in Step 6. You can't remove the account that you use to sign into Windows from here; you must do that from the Settings app in Windows itself.

Check out the list of settings you can control in Figure 11-11. It looks like a pretty thorough list, right? The Personalization option enables you to change the background image that displays in the right pane whenever there is no open message, for example, and the Signature option enables you to create an email signature — in other words, a block of text that will be appended to the bottom of every message you send automatically. Explore these settings on your own, and have fun with them!

applications

» **Using Office applications online**

» **Accessing your OneDrive storage**

» **Adding files to OneDrive**

» **Sharing a folder or file using OneDrive**

» **Creating a new OneDrive folder**

» **Using the Personal Vault**

» **Adjusting OneDrive settings**

» **Configuring online synchronization**

Chapter **12**

Working in the Cloud

Y ou may have heard the term *cloud* bandied about. The term comes from the world of computer networks, where certain functionality isn't installed on computers but resides on the network itself, in the so-called cloud.

Today, the definition of the term has broadened to include functionality that resides on the Internet. If you can get work done without using an installed piece of software — or if you store and share content online — you're working in the cloud.

In this chapter, you discover the types of applications you might use in the cloud, saving you the cost and effort of buying and installing software. In addition, I explore two Windows features that help you access your own data in the cloud: OneDrive and Sync. OneDrive is a file-sharing service that has been around for a while, but with Windows, sharing files from your computer with others or with yourself on another computer is more tightly integrated. Sync enables you to share the settings you've made in Windows on one computer with other Windows computers.

Understand Cloud-Based Applications

Certain apps, such as Maps and People, are built into Windows. You may purchase or download and install other apps, such as photo-editing apps like Photoshop or office tools such as Microsoft Word or Excel. Although these apps and applications may connect to the Internet to get information — such as updates or templates — the software itself is installed on your computer. That's the traditional way of working with software, and until recently it was the only way. However, the Internet has become fast and reliable enough to make it practical to run applications from a remote server that you connect to online.

Today, you have the option of using software in *the cloud*, meaning that you never have to install the software on your computer; instead, you simply make use of it online via your web browser. Here are some examples you can explore:

» **Online office suites:** Several companies offer online suites of business applications that include word processing, spreadsheet, and presentation apps. The most popular are Google Docs (available at docs.google.com) and Office Online (available at www.office.com). To use the Google suite you must have a Google ID (which is free). To use the Microsoft suite you must have a Microsoft ID (which you already have because you used it to sign into Windows). Figure 12-1 shows the web-based version of Microsoft Excel in Office Online, for example.

» **Email clients:** When you log into Gmail or Outlook.com, you're using email software in the cloud. In addition, many email clients can access more than one email account, such as Yahoo! and AOL.

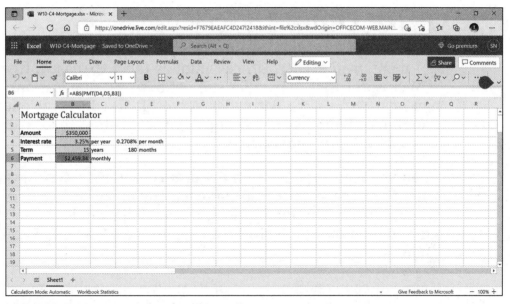

FIGURE 12-1

» **File-sharing sites:** Rather than attaching files to email messages, you're given the option of uploading and sharing them on the file-sharing site. You can do that with Microsoft's OneDrive, for example. Dropbox is another one.

» **Photo-sharing sites:** Sites such as Flickr (`www.flickr.com`) enable you to upload and download photos to them without ever installing an app on your computer. A variation on this is a site such as Viewbook (`www.viewbook.com`), where you can create an online portfolio of art samples or business presentations to share with others.

» **Financial applications:** You might use a tool such as the Morningstar Portfolio Manager (`portfolio.morningstar.com`) to maintain an online portfolio of investments and generate charts to help you keep track of financial trends. You can also use online versions of popular money-management programs, such as Intuit's free online service Mint (`www.mint.com`), through which you can access your data from any computer or mobile device.

Use Office Online

As I mentioned earlier, Office Online is the web-based, free version of Microsoft Office, a popular suite of business applications. You can sign in at `Office.com` with your Microsoft ID (the same one you use to sign into Windows itself) and have access to web-based versions of nearly a dozen applications.

When you sign in, you see a list of recently used files in the center pane (if you have any). You can click any of those to reopen the file to continue working on it. Along the left edge are icons for the various apps you can use. Point to an app to see a ScreenTip that tells its name.

Microsoft changes the layout of its online app web pages frequently, so by the time you see this, yours might not look exactly like the one in Figure 12-2, and that's okay. Just poke around, looking at the ScreenTips, to figure out what's what.

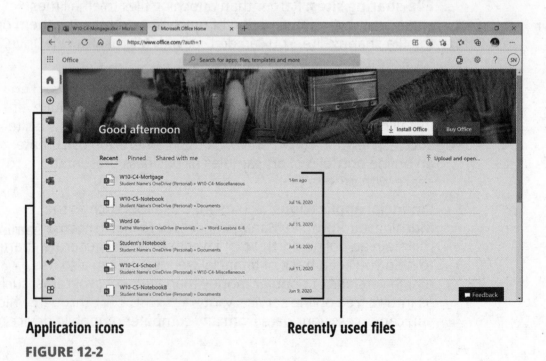

Application icons　　　　　　　　　　**Recently used files**

FIGURE 12-2

One of the icons you see along the left edge, about halfway down, is OneDrive. That's one way to get to the online OneDrive interface; another way is to go directly to `onedrive.live.com`. But I'm getting ahead of myself here — first you need to understand how OneDrive works.

Access Your OneDrive Storage

OneDrive is all about storing files and sharing them between your computer and others via the cloud. In Windows, OneDrive storage folders in File Explorer on your computer work together with the online version of OneDrive.

You can sign in to `onedrive.live.com` with your Microsoft account to work with your online OneDrive, as in Figure 12-3. You might have additional folders; Figure 12-3 just shows the basic default set. I explain the Personal Vault folder later in this chapter.

TIP

If the OneDrive online interface doesn't require to sign in, that means Windows has signed you in automatically based on your stored credentials. Nice!

You can also access your OneDrive storage from File Explorer, without having to open a browser window. In File Explorer, notice the OneDrive shortcut in the navigation bar on the left. When you click it, you see the OneDrive folders, just like you would from the web version. Figure 12-4 shows the OneDrive folders in File Explorer for the same user as in Figure 12-3.

These two seemingly different locations are actually the same location. The online location is the actual, permanent location for your OneDrive-stored content. The version that appears on your local hard drive (via File Explorer) is a mirror of the online location. A *mirror* is a copy that is kept constantly synchronized with the original. Whenever your PC is connected to the Internet, OneDrive constantly synchs itself between the two locations. When *not* connected to the Internet, you still have access to the local copy.

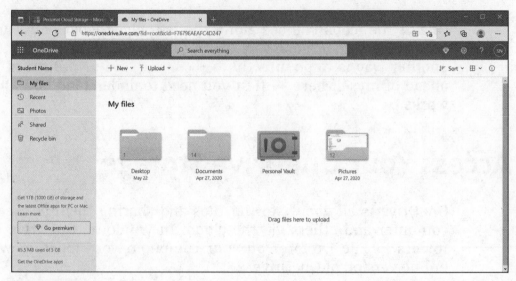

FIGURE 12-3

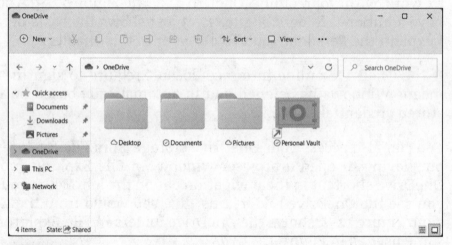

FIGURE 12-4

TIP

The first time you sign into Windows, you might be prompted to set up OneDrive. The exact prompts you see will vary (because Microsoft changes them frequently), but just follow the prompts, and you'll be set up quickly.

Because the locations are essentially the same, you can freely choose which interface you want to use to work with your OneDrive content. On your own PC, you will probably prefer the File Explorer method because it's most readily available; you don't have to open a browser window. When using someone else's PC, the web-based version is the way to go; you can access your personal files from almost any Internet-connected computing device, anywhere in the world.

Add Files to OneDrive

You can easily add or upload files to OneDrive online at any time, from mobile devices as well as your computer.

If you're on your own computer, you might want to place some of your local files on OneDrive so you will be able to access them from other computers, and so you can share them with other people online. For example, if you have a lot of family pictures, putting them on One-Drive can enable you to make them available to your entire family.

To add files to OneDrive using File Explorer:

1. Open File Explorer and navigate to the location containing the content you want to share.

Look back at Chapter 6 if you need a refresher on using File Explorer.

2. Select the content to share and then copy it to the Clipboard.

Use any method you prefer. Pressing Ctrl+C is one method.

3. Click OneDrive in the navigation pane on the left.

4. (Optional) Double-click the folder where you want to put the content (for example, Documents or Pictures) to display that location.

Alternatively, you can create a new folder.

Again, refer to Chapter 6 if you need help.

5. Paste the Clipboard content into the new location.

Use any method you prefer. Pressing Ctrl+V is one method.

Provided you are connected to the Internet, Windows will automatically and silently upload the new content to the web version of OneDrive, usually within 5–10 seconds.

If you are *not* on your own computer, and you want to upload something to your OneDrive to look at later, you must sign into the web-based version of OneDrive to do it. Follow these steps:

1. **Open your web browser and navigate to** `https://onedrive.live.com`.

   Here's a shortcut: you can leave out the live part of that address and just type onedrive.com. Your browser will figure it out and get you there.

2. **Follow the process for signing in, if prompted, using your Microsoft account username and password.**

   Your default folders appear (see Figure 12-3).

3. **Click the folder where you want to upload new content.**

   You can also create a new folder. Click New ➪ Folder, type a name for it, and click Create. Then click the new folder to move into it.

   You can return to the top level of folders at any time by clicking My Files in the navigation pane.

4. **Click the Upload button in the OneDrive toolbar and then click Files.**

   Or, if you want to upload an entire folder, click Folder.

5. **Use the Open dialog box that appears (see Figure 12-5) to locate a file or folder.**

6. **Click a file or folder.**

7. **Click Open.**

   You also can add a file to OneDrive by dragging it from a File Explorer window to an open OneDrive folder in your browser.

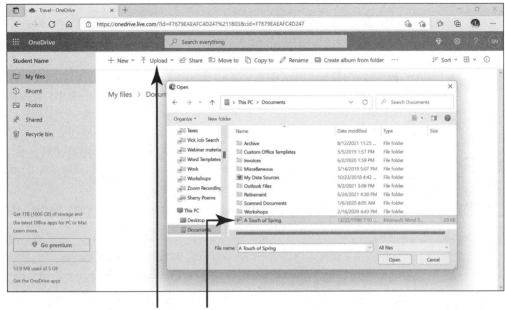

Upload button   **Select the file to upload**

FIGURE 12-5

TIP

You may want to delete a file from OneDrive, as the free storage is typically limited to 5 gigabytes (GB), though you can purchase additional storage space or get more storage if you subscribe to Microsoft 365, the subscription-based version of Microsoft Office. First, find the file that you want to delete in OneDrive. Right-click the file, and then click Delete in the shortcut menu.

While you're here, check out the other commands across the top of the OneDrive web interface window, as shown in Figure 12-5. You can move, copy, rename, and so on.

Share a Folder or File Using OneDrive

The OneDrive service not only lets you share your files with yourself as you use multiple computers, but it also lets you share files with other people. Sharing files and folders online can be easier than sending them as attachments because email apps typically limit the size of attachments that you can send at one time. You can share a file or folder that you've already stored or created by sharing a link to the file on OneDrive.

Surprisingly, the steps are very much the same whether you are using File Explorer or the web interface. The following steps handle either possibility.

To share a file or folder from OneDrive, follow these steps:

1. **Do one of the following:**

   - Open File Explorer and navigate to the location containing the file or folder you want to share.

   - Sign into the OneDrive web interface and navigate to the location containing the file or folder you want to share.

TIP

   If you want to share multiple files, it's often easier to put them in a folder and then share the folder. See the next section, "Create a New OneDrive Folder."

2. **Select the file or folder, and then right-click it. If you're using Windows 11, click Show More Options, and then click Share. If you're using Windows 10 or the web interface, just click Share.**

   The Send Link dialog box opens. See Figure 12-6.

3. **In the To box, type the email address of the person you want to share with, and then press Enter. Repeat this as many times as needed until you have added all the recipients.**

4. **(Optional) Type a note of explanation in the Message area.**

5. **(Optional) If you don't want the recipients to be able to make changes to the content, click Anyone with the Link Can Edit. Then in the Link Settings dialog box that appears, clear the Allow Editing check box and click Apply.**

6. **Click Send.**

FIGURE 12-6

Create a New OneDrive Folder

You can keep your shared files in order by placing them in folders on OneDrive. After you've placed content in folders, you can then share those folders with others. This capability to share individual folders gives you a measure of security, as you don't have to share access to your entire OneDrive content with anybody.

There are multiple methods of creating a new folder in the web interface, in Windows 10, and in Windows 11, and each has a unique combination of available options. However, here's one method that is more-or-less the same across them all:

1. **Navigate to the location where you want to create the new folder.**

2. **Right-click any empty spot in the location where you want to create the new folder.**

3. **Click New.**

A submenu appears.

4. **Click Folder.**

5. **Enter a name for the new folder.**

6. **Do one of the following:**

   - In the web interface: Click the Create button.
   - In Windows: Press Enter.

Use the Personal Vault

You might have noticed the Personal Vault folder in OneDrive. It looks like an ordinary folder, but there's more to it than that. The Personal Vault folder provides a highly secure location for you to place your most sensitive and important files. The Personal Vault requires special procedures to unlock it (including two-step verification), and it automatically locks after 20 minutes of inactivity.

To get started with the Personal Vault, click the Personal Vault icon, either in File Explorer or the web-based OneDrive interface. Then follow the prompts to set it up.

WARNING

There's one catch to the Personal Vault: by default you can store only three files in it. If you want to be able to store more files there, you must upgrade your OneDrive account to a Premium (paid) one. You can get a OneDrive-only Premium account for about $2 a month (what a deal!), or a Microsoft 365 account that also gives you a subscription to the full Microsoft Office applications. With a Premium account you can store an unlimited number of files in the vault, up to 1TB.

TIP

You can change the default time delay on the automatic lock from 20 minutes to something longer if you like. Open the OneDrive settings dialog box (covered in the next section) and change the setting there.

Adjust OneDrive Settings

OneDrive has a variety of options you can adjust to tweak how it works. To access them, follow these steps:

1. In the notification area (the small icons to the left of the clock in the lower right corner of the Windows desktop), locate the OneDrive icon. It looks like a cloud.

2. Right-click the OneDrive icon and click Help & Settings.

A menu appears.

3. Click Settings.

The Microsoft OneDrive dialog box opens. It contains several tabs, each controlling a different type of settings. See Figure 12-7.

FIGURE 12-7

4. **Click each tab of the dialog box and check out the available settings:**

Settings tab: Many checkboxes that toggle various features on and off.

Account tab: Add another account, choose the folders that should sync automatically with OneDrive, and adjust the personal vault's lock timer.

Backup tab: Sets up an automatic backup of the files in your Documents, Pictures, and Desktop folders.

Network tab: Enables you to adjust the upload and download rates on your network. (This is a techie thing; most people can ignore it.)

Office tab: Enables Office applications to synch Office files, and lets you decide how to handle synchronization conflicts.

About tab: Provides information about the OneDrive app.

5. **Click OK to close the dialog box when you are finished looking around.**

Configure Online Synchronization

Did you know that your Windows settings can follow you wherever you go? It's true! When the Sync feature is enabled in Windows, then whatever PC you sign into with the same Microsoft account will know your personalization preferences, such as your preferred desktop background. Your personal files from your OneDrive storage will be available automatically, too.

To sync, the Sync feature must be turned on in Settings, which it is by default. If, for some reason, it's been turned off, you must turn it on.

To enable the Sync feature (if it's not already enabled), follow these steps:

Windows 11:

1. **Open the Settings app.**
2. **Click Accounts.**

3. **Click Windows Backup.**

4. **Make sure the Remember My Preferences slider is set to On.**

5. **Click Remember My Preferences to display a set of check boxes for individual options. See Figure 12-8.**

6. **Close the Settings app.**

FIGURE 12-8

Windows 10:

1. **Open the Settings app.**

2. **Click Accounts.**

3. **Click Sync Your Settings.**

   The Sync Your Settings screen appears. See Figure 12-9.

4. **Turn each option on or off as desired.**

5. **Close the Settings app.**

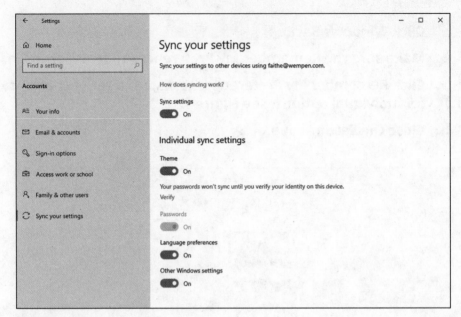

FIGURE 12-9

Syncing works only with Windows 10 settings and settings for apps that you buy from the Windows Store.

TIP

IN THIS CHAPTER

» **Using discussion boards and blogs**

» **Participating in a chat**

» **Understanding instant messages**

» **Exploring Teams**

» **Exploring Skype**

» **Using a webcam**

» **Getting an overview of collaborative and social networking sites**

» **Signing up for a social networking service**

» **Understanding how online dating works**

Chapter **13**

Connecting with People Online

The Internet offers many options for connecting with people and sharing information. You'll find discussion boards, blogs, and chat on a wide variety of sites, from news sites to recipe sites, sites focused on grief and health issues, and sites that host political- or consumer-oriented discussions. There are some great senior chat rooms for making friends, and many sites allow you to create new chat rooms on topics at any time.

Today's Internet goes far beyond just typing text on a website, though. A variety of real-time messaging apps such as Microsoft Teams and Zoom enable you to talk to people via video and/or audio chat. And text-centric chat programs such as Discord provide a conversation space for real-time text and audio chats. Social networking services like Facebook also provide their own communication services.

In this chapter, I look at some ways of sharing information and connecting with others, and tell you how to do so safely.

As with any site where users share information, such as social networks and blogs, you can stay safer if you know how to sidestep some abuses, including **data mining** (gathering your personal information for commercial or criminal intent), **social engineering** ploys that try to gain your trust and access to your money, ID theft scams, and so forth. If you're careful to protect your privacy, you can enjoy socializing online without worry.

Use Discussion Boards and Blogs

A **discussion board** is a place where you can post written messages, pictures, and videos on a topic such as home improvement. Others can reply to you, and you can respond to their postings.

In a variation on discussion boards, you'll find **blogs** (web logs) everywhere you turn. Blogs are websites where the owner can post multiple articles on a topic such as cooking or politics, and the articles appear with the most recent ones first. In that sense, it really is a "web log," in that it is a log of someone's thoughts and ideas over time. Blogs are not just for individuals, though; companies also have blogs where they post information about their product or service. People who read blogs can also post comments about those blog entries.

Discussion boards and blogs are **asynchronous**, which means that you post a message (just as you might on a bulletin board at the grocery store) and wait for a response. Somebody might read it that hour — or ten days or several weeks after you make the posting. In other words, the response isn't instantaneous, and the message isn't usually directed to a specific individual.

You can find a discussion board or blog about darn-near every topic under the sun, and these are tremendously helpful when you're looking for answers. They're also a great way to share your expertise — whether you chime in on how to remove an ink stain, provide historical trivia about button styles on military uniforms, or announce the latest breakthroughs in your given field. Postings are likely to stay up on the site for years for people to reference.

To try using a discussion board, follow these steps:

1. **Enter this URL in your browser's Address box and press Enter:** `http://answers.microsoft.com/en-us.`

   Some discussion boards require that you become a member, with a username, and sign in before you can post. This site lets you sign in with the same Microsoft account that you use to sign into Windows.

2. **Scroll down to the bottom of the page. Notice in the lower-left corner of the page that the default language and region is English. You can click that link and then click another language of your choice.**

3. **Scroll back up and click a topic area under Browse the Categories, such as Windows.**

4. **In the topic list that appears, scroll down to the All section and then click on a topic to read about it.**

   Figure 13-1 shows an example.

   When you click a posting that has replies, you'll see that the replies are listed down the page in easy-to-follow threads, which arrange postings and replies in an organized structure. You can review the various participants' comments as they add their ideas to the conversation.

5. **To reply to a posting yourself, first click the posting and then click the Reply link (or Comment link, depending on the message type). For this site, you then type your comments in the Reply (or Comment) box, scroll down, and click Submit.**

   Figure 13-2 shows a reply being composed.

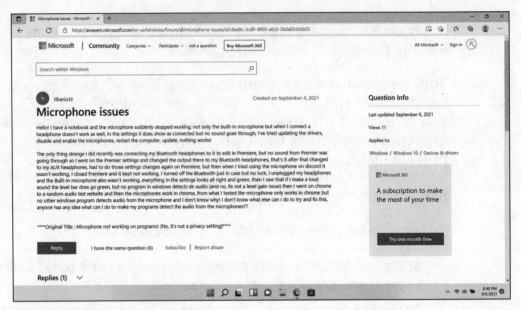

FIGURE 13-1

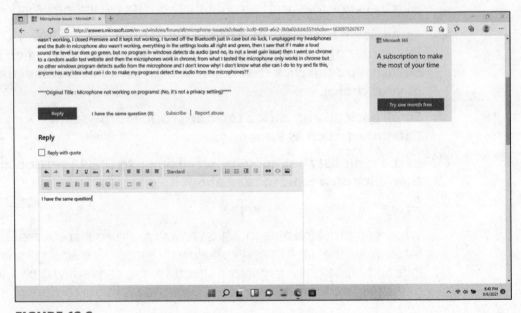

FIGURE 13-2

WARNING

The first time you reply to something, you may be prompted to enter a display name. Don't use your full first and last name; your first name and last initial is better because it safeguards your privacy. Or better yet, use a nickname or alias.

Participate in a Chat

A **chat room** is an online space where groups of people can talk back and forth via text, audio, web camera, or a combination of media. Figure 13-3 shows some of the popular free chatrooms at a site called Paltalk.com, for example. (That's not a recommendation for it; I haven't used it enough to evaluate it.)

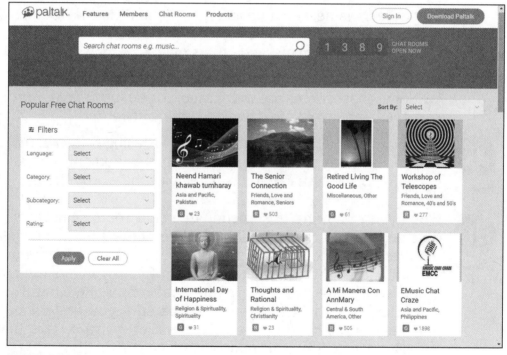

FIGURE 13-3

In chat, you're having a conversation with one or more people in real time, and your entire conversation appears in the chat window. You may have participated in a chat with a customer service rep while browsing in an online store, for example. Here are some characteristics of chat that you should know about:

>> When the chat is over, unless you save a copy, the conversation is typically gone.

>> Interactions are in real time (synchronous), which means you can interact with others in the moment.

>> Several people can interact at once on some sites, although this can take getting used to as you try to follow what others are saying and jump in with your own comments.

>> When you find a chat you want to participate in, sign up to get a screen name, and then you simply enter the chat room, type your message, and submit it. Your message shows up in the stream of comments, and others may — or may not — reply to it.

>> Some chat services are web-based, so you just use your browser; others require you to download their software.

WARNING

Be cautious when downloading and installing software; make sure the site you are getting it from has a good online reputation. Do a web search for its name plus the word *reviews.*

WARNING

When you're talking to someone in a chat room that hosts multiple people, you might be able to, if you'd like, invite one person to enter a private chat room, which keeps the rest of the folks who wandered into the chat room out of your conversation. Also, others can invite you into private chat rooms. Be careful whom you interact with in this way, and be sure you understand the motivations for making your conversation private. This may be entirely reasonable, or it may be that you're dealing with someone with suspect motivations.

Before you get started, check out the website's Terms of Use and privacy, monitoring, and abuse-reporting procedures to understand the safety protections that are in place. Some sites are well monitored for signs of abusive content or interactions; others have no monitoring at all. If you don't like the terms, find a different site.

Understand Instant Messages

Instant messages (often called just **IMs**) used to be referred to as real-time email. It used to be strictly synchronous, meaning that two (or more) parties could communicate in real time, without any delay. It still can be synchronous, but now you can also leave a message that the recipient can pick up later.

Instant messaging is a great way to stay in touch with younger generations who rarely use email. IM is ideal for quick, little messages where you just want an answer without writing an email, as well as for touching base and saying hi. Text messaging on cellphones is largely the same phenomena: This isn't a tool you'd typically use for a long, meaningful conversation, but it's great for quick exchanges.

Depending on the IM service you use, you can do the following:

>> Write notes to friends, grandchildren, or whomever

>> Talk as if you were on the phone

>> Send photos, videos, and other files

>> Use little graphical images, called *emoticons* (such as smiley faces or winks), *avatars* (characters that represent you), and *stickers* (cartoony pictures and characters), to add fun to your IM messages

>> See participants via web cameras

>> Get and send email

» Search the web, find others' physical location using Global Positioning System (GPS) technology, listen to music, watch videos, play games, bid on auctions, find dates, and more

» Track the history of conversations and even save transcripts of them to review later

Instant messaging programs vary somewhat, and you have several to choose from, including the T app that's included in Windows 10 and the Teams app that's included in Windows 11. Other messaging apps include Facebook Messenger and Discord.

TIP

IM is one place where people use shortcut text. Some of this will be familiar to you, such as *FYI* (for your information) and *ASAP* (as soon as possible). Other short text may be less familiar such as *AFAIK* (as far as I know). Visit pc.net/slang for a table of common shortcut text terms. Knowing these will make communicating with younger folks more fun.

WARNING

Consider what you're saying and sharing in IM and how you'd feel if the information were made public. IM allows you to store your conversation history, which is super-useful if you need to go back and check something that was said, but it has its downside. Anything you include in an IM can be forwarded to others. If you're at work, keep in mind that many employers monitor IM (and email) conversations.

WARNING

If you run across illegal content — such as child pornography — downloading or continuing to view this for any reason is illegal. Report the incident to law enforcement immediately.

WARNING

You can send IMs from a computer to a mobile phone (and vice versa) and from one mobile phone to another. If you include your mobile phone number as part of your IM profile, anyone who can see your profile can view it. This is useful information for both friends and criminals, so it's important to consider whether you want your number exposed — especially if you have many people on your Contacts list whom you don't personally know.

Explore Microsoft Teams

The Teams app comes with Windows 11. It enables you to send IMs to friends, as well as to make voice and video calls and create groups of people who will share photos and videos.

The first time you open the Teams app, you are led through a setup process where you enter your email address (that is, your Microsoft ID) and a phone number. Work through the prompts, accepting the defaults when you don't know what to pick.

Your contacts, if you have any yet, appear on the left side of the screen. You can double-click one of your contacts to start a text chat with that person. If they don't have a Microsoft account associated with them, but they do have a phone number, Teams will send them an invitation via SMS text message (that is, a text message to their phone).

To add a contact to Teams (so you can chat and share files with that person), follow these steps

1. **Click Invite to Teams.**

   A box appears with a hyperlink in it. To invite someone, you'll share that hyperlink with them.

2. **Click Copy to copy the hyperlink to the Windows Clipboard.**

3. **Open your email application and start composing a new message to that person.**

4. **In the body of the message, type an explanation of what you are sending them and why. Then press Ctrl+V to paste the link. Figure 13-4 shows a message ready to send. (The address of your URL will be different.)**

5. **Finish sending the message as you normally would.**

   When the person gets your message and clicks on the link, it adds you to their contact list.

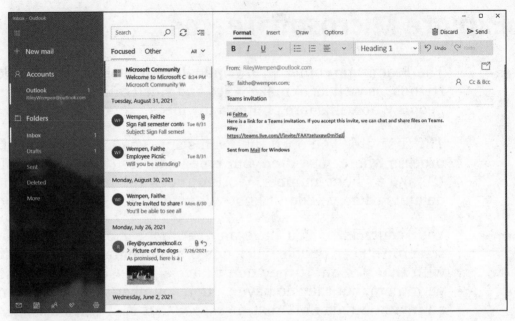

FIGURE 13-4

6. **When someone sends you a message, it appears in your contact list, appearing in bold and at the top so you will notice it more easily.**

The first time someone sends you a message, a prompt appears asking you to Accept or Block. If you choose Accept, that person's message will come through to you, as will all future messages from them. Figure 13-5 shows a conversation happening.

From that point on you can have a text-based conversation with the person, or you can have a voice or video call with them. To have a video call, click the Video call icon in the upper-right corner of the screen. To have a voice call, click the Audio call icon. See Figure 13-5.

To add more people to the chat, for a group experience, click the Add People icon and follow the prompts.

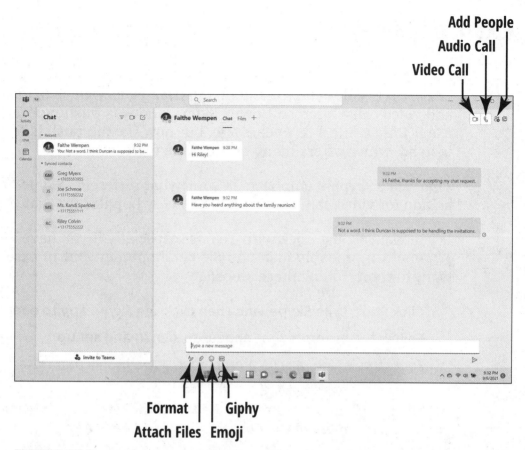

Add People
Audio Call
Video Call

Format   **Giphy**
Attach Files **Emoji**

FIGURE 13-5

There are four icons below the chat window, each of which adds a special functionality to your chat. From left to right:

» **Format:** Opens a pane where you can format the text you are typing, giving it different fonts, sizes, colors, and so on.

» **Attach files:** Enables you to upload a file from your computer as an attachment you can send to your chat-mate.

» **Emoji:** Provides a palette of yellow smiley faces (actually a whole gamut of emotions of faces) that you can use to emphasize your meaning.

» **Giphy:** Opens a pane where you can type keywords and find the perfect animated graphic that expresses your meaning.

Explore Skype

The Skype app comes with both Windows 10 and Windows 11. Like Teams, it enables you to send IMs to friends, as well as to make voice and video calls. Microsoft owns Skype, and was pushing everyone to use it for a while a few years ago, but now the big push is for Teams instead, which offers many of the same features.

That said, Skype's emphasis is somewhat different. Skype began as an app for video calling, and remains very popular for that.

If you want to just stick with Teams (especially if you have Windows 11), you can probably just skip Skype altogether. But in case you do want to try it, follow these steps:

1. **Click Start, type Skype, and then click the Skype app to open it.**

2. **Follow the prompts that appear to sign in and set up.**

   For example, you may be prompted to add a profile picture, test your audio and video, and make a free test call.

3. **Click the Contacts button at the top of the left pane. Existing contacts appear in the Contacts list if you have any.**

   You can invite new contacts not in your Contacts list by doing the following:

   a. *Click New Contact and click Invite to Skype.*

   b. *In the Share and Connect dialog box, scroll down to the bottom and click Email. The Mail app (or your default email app) opens and starts composing a new message with the hyperlink in it.*

   c. *Fill in the recipient's email address in the To box and click Send.*

4. **The first time someone sends you a message, a prompt appears asking you to Accept or Block. If you choose Accept, that person's message will come through to you, as will all future messages from them. Figure 13-6 shows a conversation happening.**

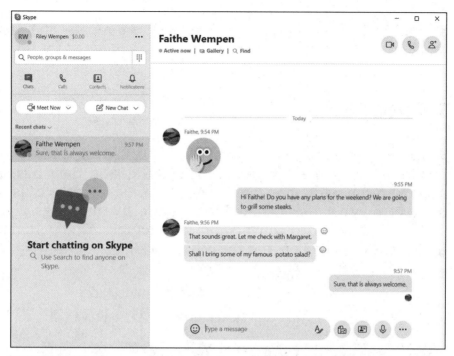

FIGURE 13-6

If that chat interface in Figure 13-6 looks familiar, it's because it's a whole lot like Teams. And just like in Teams, you can use the buttons at the top right corner of the screen to start a Video call or an Audio call or create a new group.

The buttons to the right of the chat message typing area are:

» **Open Richtext Editor:** This enables you to format your text.

» **Add Files:** Use this to share a file with your chat-mate.

» **Send Contacts to This Chat:** Lets you attach contact information.

» **Record a message:** Lets you use your PC's microphone to record an audio message that is then sent like a file. The recipient can click the file to hear the message.

» **More:** Opens a menu of additional options, including Location, Video Message, Schedule a Call, Create Poll, and OneDrive.

TIP

Skype and Teams are both also available as a mobile app. Install them on a smartphone or tablet to use those devices for IM.

Use a Webcam

Webcams are relatively inexpensive, and laptops and most computer monitors now come with webcams embedded in their lids. You can use a webcam with apps like Teams or Skype to make calls over the Internet, or other apps to have face-to-face, live meetings. Figure 13-7 shows a Teams video conversation in progress. One participant has their camera turned on and the other doesn't.

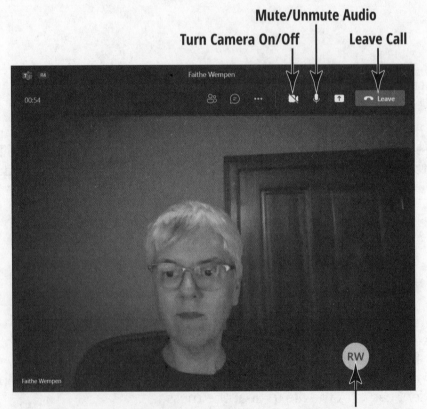

FIGURE 13-7

WARNING

A webcam can be a great way to communicate with friends and family, but it can quickly become risky when you use it for conversations with strangers.

>> Giving your image away, especially one that may show your emotional reactions to a stranger's statements in real time, simply reveals too much information that can put you at risk.

>> If you use a webcam to meet online with someone you don't know, that person may expose you to behavior you'd rather not see.

>> Note that webcams can also be hijacked and turned on remotely. This allows predators to view and listen to individuals without their knowledge. When you aren't using your webcam, consider turning it off or disconnecting it if it isn't a built-in model.

WARNING

Teens in particular struggle to use good judgment when using webcams. If you have grandchildren or other children in your care, realize that normal inhibitions seem to fall away when they aren't physically present with the person they're speaking to — and many expose themselves, figuratively and literally. In addition to having a conversation about appropriate webcam use with children and teens, it may be wise to limit access to webcams.

Get an Overview of Collaborative and Social Networking Sites

Although you may think kids are the only active group using social networking, that isn't the case. In fact, people 35–54 years old make up a large segment of social networkers.

There are several types of sites where people collaborate or communicate socially. The following definitions may be useful:

» **Wiki:** A website that allows anyone visiting to contribute (add, edit, or remove) content. Wikipedia, for example, is a virtual encyclopedia built by users providing information in their areas of expertise. Because of the ease of collaboration, wikis are often used when developing group projects or sharing information collaboratively.

» **Blog:** An online journal (*blog* is short for *web log*) that may be entirely private, open to select friends or family, or available to the public. You can usually adjust your blog settings to restrict visitors from commenting on your blog entries, if you'd like.

» **Social networking site:** This type of website allows people to build and maintain a web page and create networks of people that they're somehow connected to — their friends, work associates, and/or other members with similar interests. Most social networking sites also host blogs and have social networking functions that allow people to view information about others and contact each other. Facebook is one of the most popular, shown in Figure 13-8.

» **Social journaling site:** Sites such as Twitter allow people to post short notes online, notes that are typically about what they're doing at the moment. Many companies and celebrities are now *tweeting,* as posting comments on Twitter is referred to. You can follow individuals on Twitter so you're always informed if somebody you're a fan of makes a post.

Followers often comment on each other's tweets with another short message. Tweets have a limit of 140 characters so tweets and responses to them must be concise.

FIGURE 13-8

Sign Up for a Social Networking Service

Many social networking sites, such as Facebook or Pinterest, are general in nature and attract a wide variety of users. Facebook, which was begun by some students at Harvard, has become today's most popular general site, and many seniors use its features to blog, exchange virtual gifts, and post photos. Other social networking sites revolve around particular interests or age groups.

When signing up for a service, understand what is *required* information and what is optional. You should clearly understand why a web service needs any of your personally identifiable information and how it may use that information — before providing it. Carefully consider the questions that sites ask users to answer in creating a profile.

TIP

Accepting a social networking service's default settings may expose more information than you intend.

Walk through the signup process for Facebook to see the kinds of information it asks for. Follow these instructions to do so:

1. **Type this URL into your browser's address line:** www.facebook.com. **At the sign-in prompt, click Create New Account.**

2. **In the sign-up form that appears (see Figure 13-9), enter your name, email address, a password, your gender, and birthdate.**

Note that the site requires your birthdate to verify that you are old enough to use the service, but you can choose to hide this information from others later if you don't want it displayed. (I recommend hiding your birthdate to avoid senior-specific scams.)

3. **Click the Sign Up button. The website will send a text to the phone number you entered containing a confirmation code. Enter the code on the website to prove that it is really your phone number.**

4. **Follow any additional prompts that appear.**

You now have a Facebook account, and can continue to fill out profile information for Facebook on the following screens, clicking Save and Continue between screens. You can sign into Facebook at any time and view your personal home page, add friends, post messages, or view others' home pages and postings.

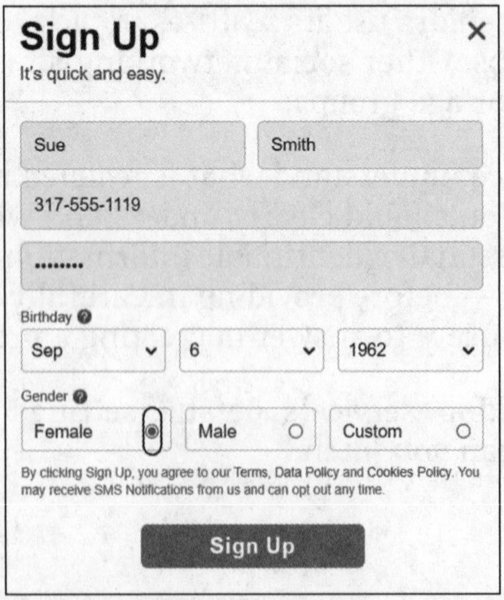

Sign Up
It's quick and easy.

Sue | Smith

317-555-1119

••••••••

Birthday
Sep | 6 | 1962

Gender
Female ⦿ | Male ○ | Custom ○

By clicking Sign Up, you agree to our Terms, Data Policy and Cookies Policy. You may receive SMS Notifications from us and can opt out any time.

Sign Up

FIGURE 13-9

TIP

Remember that social networking sites sometimes ask for information during signup that they use to provide you with a customized experience that suits your needs. But sometimes the information isn't needed at all for the service they're providing you — they simply want it for marketing purposes, to show to other members, or to sell.

TIP

It's often very difficult to remove information from sites if you later regret the amount of information you've shared. It's best to be conservative in the information you share during the sign-up process; you can always add more later.

Understand How Online Dating Works

Many seniors are making connections with others via online dating services, and if you've been wondering if this route could be for you, here's how you can jump into the world of online dating:

>> Choose a reputable dating site.

>> Sign up and provide information about your likes, dislikes, preferences, and so on. This often takes the form of a self-guided interview process.

>> Create and modify your profile to both avoid exposing too much personal information and ensure that you're sending the right message about yourself to prospective dates.

>> Use search features on the site to find people who interest you or other people who match your profile. Send them messages or invitations to view your profile.

>> You'll get messages from other members of the site, to which you can respond (or not). Use the site's chat and email features to interact with potential dates. You may also be able to read comments about the person from others who've dated them, if the site has that feature.

>> When you're comfortable with the person and feel there might be a spark, decide if you want to meet the person offline.

TIP

Formal dating sites aren't the only places where people meet online, but they typically have the best safeguards in place. If you want to interact with people you meet on other sites, you should provide your own safeguards. Create a separate email account (so you can remain anonymous and abandon the email address if needed). Many dating sites screen participants and provide strong reporting measures that are missing on other types of social sites, so be particularly careful. Take your time getting to know someone first before connecting and always meet in a public setting.

Select your online dating service carefully.

» Look for an established, popular site with plenty of members and a philosophy that matches your own.

» Review the site's policy regarding your privacy and its procedures for screening members. Make sure you're comfortable with those policies.

» Use a service that provides an email system (sometimes called *private messaging*) that you use for contacting other members. By using the site's email rather than your own email address, you can maintain your privacy.

» Some sites, such as www.eharmony.com/senior-dating, offer stronger levels of authenticating members, such as screening to make you more confident that you know with whom you're interacting.

» Visit a site such as www.consumer-rankings.com/dating for comparisons of sites. Whether you choose a senior-specific dating site such as Ourtime.com or DatingForSeniors.com, or a general-population site such as PerfectMatch.com, reading reviews about them ahead of time will help you make the best choice.

TIP

If you try a site and experience an unpleasant incident involving another member, report it and make sure the service follows through to enforce its policies. If it doesn't, find another service.

4

Having Fun

IN THIS PART . . .

Finding and playing computer games

Creating and viewing photos and videos

Listening to music on your PC

Chapter **14**

Let's Play a Game!

E veryone loves a good game, right? You might be surprised to find out just how much time even serious computer users spend unwinding with computer games. We might not all agree on what constitutes a *good* game, but thankfully there's a computer game for every taste, from "Let's find all the lollipops!" to "Kill everything in sight!"

There are so many games available today that it can be hard to wade through all the choices to find something you really enjoy. So, this chapter takes you on a whirlwind tour of the various genres available and how to get them.

Learn the Types of Game Delivery

Games are available in so many different formats that it can be bewildering. Here are some of the options:

» **Download:** This is the modern method of acquiring a game that you install on your local PC. You download the Setup utility and run it to install the game.

- » **On CD or DVD:** This is rather old-school, since few computers these days have optical drives, but you can still buy a CD or DVD that contains a Setup utility for a game. You run that Setup utility to install the game on your local PC. Most games that come on optical discs don't require the Internet to play them, but some have extra features you can access when online.

- » **Single-game website:** With this type of game, you don't have to download anything. You simply visit a certain website using your browser and play the game on that website.

- » **Social media:** Some social media sites have games you can play while signed into that site. Facebook in particular has hundreds of them.

- » **Multi-game website:** These websites exist just to offer online games. Some are free, at least for the basic games; some allow you more games, more time, or more privileges if you pay a small amount each month.

- » **Mobile app:** Both the Apple Store and the Google Play Store offer game downloads, either for free or for small payments. After you buy a game in one of these stores, you can install it on any of your mobile devices, provided you sign into the device using the same user ID that you used to purchase it.

- » **Game console:** Game console systems such as PlayStation, Xbox, and Nintendo Switch all offer many hundreds of games you can download by connecting your console to the Internet. Some console systems (including all the older ones) also enable you to buy game cartridges or discs and insert them into the console to play games.

Which is the best? That really depends on your situation. If you have kids visiting your home, they will probably tell you which gaming console they like.

But if you're shopping just for yourself? Personally, I like simple social media and website games because they're free (with optional in-app purchases) and because they're convenient — I can play them wherever I might be, on a variety of devices. I play a few games on my phone too, like Boggle with Friends. But some games are too big and complicated to deliver that way, like World of Warcraft and Diablo 4, so I end up downloading and installing them.

Explore the Various Gaming Genres

No matter what you're interested in, or what constitutes entertainment for you, there's a game category that came into being specifically because of people like you. Think I'm exaggerating? Google your top two favorite things in the world plus the word "game" and see what you get.

When considering games, it is helpful to think about them in categories. Here are some of the most popular ones.

REMEMBER

I don't provide hyperlinks for the specific games here because addresses tend to change over time, but you can easily find any game mentioned here by searching for it using a web search engine. Most of the screenshots in this chapter are from Facebook versions of the games, but there are plenty of non–Facebook versions as well:

» **Casual games:** A casual game is one that isn't very complicated. You can jump in and out of a casual game for a few minutes at a time without stressing out. The penalty for losing is low; usually you just have to re-do whatever level you failed to complete. Some examples of casual games include the time management games like Diner Dash, match-three type of games like Candy Crush (see Figure 14-1), and hidden object games like Criminal Case. The latter two are very popular on Facebook; you can play them for free there.

» **Word and number games:** Word games involve manipulating letters and words in various ways. There are online versions of popular physical games like Boggle and Scrabble (and their equivalents, like Words with Friends, shown in Figure 14-2), as well as every kind of puzzle you ever solved in a magazine, like word searches, anagrams, and logic problems. There are also number games like Sudoku.

FIGURE 14-1

FIGURE 14-2

» **Card games:** Do you like to play cards, but you don't have anyone local who plays the games you do? Problem solved. Card game apps are available for almost every card game you can imagine, and you can play against random players from all over the world, your friends and relatives, or even the computer itself. There's Solitaire, Uno, Bridge, Canasta, Cribbage, Phase 10, and more. Figure 14-3 shows an Uno game on Facebook.

FIGURE 14-3

» **Board games:** Many favorite board games are available as computer apps. Because of trademarks and licensing issues, board games can be harder to find online than card games, but versions are out there if you look. For example, you can play Monopoly on Monopoly Plus for the PC. And that's just one game — look for all your favorites.

» **Casino games:** If you love the excitement of going to a casino and playing table games and slot machines, you'll enjoy computer casino games that simulate that experience. Although some sites allow you to play for real money, I don't recommend them; stick with the free sites where you can play with "pretend money." One popular offering is DoubleDown Casino, accessed from Facebook. See Figure 14-4.

FIGURE 14-4

» **Shooter games:** For some reason, a lot of people seem to love to shoot things. A huge variety of shooting games cater to that. What can you shoot? Just about anything, from fictitious people (Grand Theft Auto) to zombies and demons (Diablo 4). There are even hunting and safari games where you can shoot animals.

Shooting games where what you see on the screen is from the perspective of the character you are playing are called **first-person shooters**. Games where you observe and control the character from a distance are **third-person shooters**. Diablo 4, shown in Figure 14-5, is a third-person shooter. Half-Life 2, shown in Figure 14-6, is a first-person shooter.

» **World building:** Are you more of a builder than a destroyer? Then you might enjoy games that enable you to create cities, civilizations, and even entire planets. SimCity is a perennial favorite, in which you build and manage cities. Many online games like Elvenar and Taonga let you build fantasy civilizations, harvest and sell resources, negotiate war or peace with nearby civilizations, and more.

FIGURE 14-5

FIGURE 14-6

>> **Simulations:** Whatever you enjoy doing — or fantasize about doing — there's probably a game that lets you play at doing it. There are games that simulate farming, managing a team, giving people makeovers, running a casino, building and operating a railroad, and so many more activities. One of the most popular simulation games of all time, The Sims (Figure 14-7), is a simulation of life itself — you create characters, build homes for them, and then manage their lives from birth to old age.

>> **Sports and driving:** Sports fans of all kinds enjoy apps that let them put their knowledge of sports to the test. There are games for nearly every sport you've ever heard of: football (American and international — also known as soccer), baseball, bowling, cricket, auto racing, horse racing, flying airplanes . . . you name it.

>> **Fantasy and adventures:** Fans of science fiction and fantasy novels often gravitate toward adventuring games where they can play the role of a hero on a quest. Many of these games are full-featured and immersive, allowing the player to slip into the alternate reality of an imaginary world. Winning the game might involve a lot of shooting and fighting, but it also might involve puzzle solving, finding your way through mazes, and negotiating deals with other characters.

FIGURE 14-7

That might seem like a lot of different game categories, but this list barely scratches the surface.

Understand How Game Makers Get Paid

A lot of games are free — even some of the really good ones. So perhaps you're wondering "How is that possible?" How can companies afford to spend thousands of hours creating a game and then give it away? To answer that question, I have to take you back in time to explain how software has traditionally been sold.

In the early days of personal computing, apps were a lot smaller and easier for a single individual to write than they are today. Individual software developers often distributed their apps (especially games) for free on an honor system, where if you liked the game and wanted to keep it, you were supposed to send the maker money. This system, called **shareware**, enabled a lot of small-time programmers to get their work out there to the public. An alternative called **freeware** gave the software away for free, but kept the copyright on it, and specified that nobody else could charge money for it. **Public domain** software was one step beyond that, where the owner legally gave up all rights to say what others could do with it. All three of those licensing types are still out there today, although they're not as common as they used to be.

As computers became more capable, apps became more complex to create, and large teams of programmers worked together to create retail games that large companies loaded onto CDs and DVDs and sold to the retail-buying public. With this model, you would buy a game on a disc, install it on your PC, and it was yours to enjoy forever (or at least until you got bored with it or a new version came out that was better).

Then the mobile app thing happened. Mobile devices need small, simple apps because they don't have a lot of storage space, and developers obliged, harkening back to the early days when a single programmer could write an entire program. App stores like the Apple Store and Google Play became distribution hubs for their products.

Consumers expected most apps to be free or extremely inexpensive (a few dollars) for mobile apps, so software makers had to figure out a way to earn a living from their work.

Most apps today are free to download and use but earn money in one of two ways. The first is to sell in-game ad space. Ads pop up as you play, usually between rounds in a game, and the game owner makes money from ad sales. Some games allow you to pay a one-time fee to remove the ads. The second method is to offer the basic version of a game for free, and then charge small amounts within the game for extras. This system is sometimes called "free to play; pay to win" because as you progress through the game, certain activities are difficult to succeed at without some special boosters, weapons, or whatever that you can buy for a few dollars charged to a stored credit card. For example, Figure 14-8 shows a game called Taonga inviting me to buy some in-game goodies for the low price of $1.99.

FIGURE 14-8

A Few of My Favorites

As you explore the world of gaming, you will make your own favorites list, and it probably won't look anything like mine! But here are a few games and gaming sites that I have enjoyed and spent a lot of time with, and maybe some of them will appeal to you as well.

» **The Sims:** I showed you this one back in Figure 14-7. It's a relaxing, stress-free simulation game that runs on your local PC. You create human-like characters, put them in houses and jobs, and then micro-manage their lives to make sure they succeed. If you've ever fantasized about micro-managing your kids' lives (or grandkids) so they won't keep messing things up, you'll love this.

» **Taonga:** A classic online world-building game. Here's the premise: you're on an island that is lush with vegetation, and your job is to create your own tropical island paradise there by chopping down trees to build structures, raising crops and planting fruit trees, raising cows and chickens, and interacting with the friendly locals. The web-based version is `https://taongafarm.com` (shown in Figure 14-9), and you can also play it from Facebook. Free, with optional in-game purchases.

FIGURE 14-9

» **Boggle with Friends:** This game is only on Apple and Android mobile devices; look for it in the app store for your phone or tablet. It's just like the regular Boggle game, where you try to find words in a grid of cubes with letters on them. Figure 14-10 shows a screenshot from my phone.

FIGURE 14-10

» **BigFish Games:** This isn't a single game, but a website that sells older casual games that run on your local PC. Scoff if you want at the fact that most of the games here are from 15+ years ago, but a lot of them are fun and addictive, and you get a one-hour free trial of each game before committing to buy it. Some of my favorites from this site include Out of Your Mind, Miss Management, and Fairy Godmother Tycoon. Check it out at `https://www.bigfishgames.com`.

» **Pinball FX3:** Here's a great pinball simulator from Microsoft, available from the Microsoft Store app (`www.microsoft.com/en-us/p/pinball-fx3/9pckbvf3p67h`). You can buy simulations of lots of old pinball machines that you might remember from decades past and hanging out in places where pinball machines tended to be. Figure 14-11 shows a simulation of Funhouse, a machine I have fond memories of.

FIGURE 14-11

» **Puzzle Baron:** Not a specific game, but a website where you can work dozens of different kinds of word and number puzzles online. This is my go-to site when I want to do a logic puzzle or two. Find it at `https://www.puzzlebaron.com/portfolio-category/websites/`.

» **A Dark Room:** This text-centric free online adventure starts out as a mystery. You wake up in a dark room. What do you do next? Start clicking on things to find out what's going on. I won't say more; it'll spoil the surprise and fun of discovering things for yourself (`http://adarkroom.doublespeakgames.com`).

» **The Paperclip Game:** I have to start out by saying this game is completely pointless and silly. In fact, it was created as a joke, a spoof on the pointlessness of capitalism. And yet you may find yourself up until 3 a.m. playing it. Like A Dark Room, it starts out with no instructions, and you must blindly click your way around until you figure out what's going on. When I finally reached the end, I thought to myself "Well, that was a complete waste of time!" And then I played it again (`https://www.decisionproblem.com/paperclips/index2.html`).

with the Camera app

» Capturing audio recordings with Voice Recorder

» Playing videos with the Movies & TV app

» Transferring photos and videos from a phone or digital camera

» Viewing and editing photos in the Photos app

» Creating your own videos with Video Editor

Chapter **15**

Creating and Viewing Digital Photos and Videos

The world has discovered that it's fun and easy to share photos and videos online, and that's probably why everybody is in on the digital image craze. Most people today have access to a digital camera (even if only on their cellphones) and have started manipulating and swapping photos and videos like crazy, both online and off.

But today, your phone, tablet, and computer not only let you upload and view pictures and videos: You can use a built-in camera to take your own pictures or record videos and play them back. You can also buy videos (movies and TV shows, for example) and play them on your computer or other device, such as a tablet.

Windows offers several useful applications for working with photos and video. In this chapter, you learn about them, as well as how to set up your computer's microphone and camera to create your own photos and video content.

Capture Pictures and Video with the Camera App

Most laptops come with a built-in camera that faces the user. This enables you to have video chats with people and participate in video meetings. Most desktop PCs do not have a built-in camera; instead you install an external camera that connects to a USB port on the computer. Both are referred to as a **webcam**, which is short for web camera.

TIP

Even if you have a built-in camera on your laptop, you can still add an external USB webcam. An external webcam can be pointed in any direction, so you can film more than just yourself using the computer.

A webcam can capture still images (photos), and it can also capture video. You can save the video to a file, making your own mini-movie, or you can use a streaming application to transfer the video to someone else's computer over the Internet.

A built-in webcam should work automatically with no setup required. An external webcam usually just needs to be plugged into a USB port and it starts working automatically. If it doesn't, check the documentation that came with the camera to see if there is anything special you need to do, like download and install a driver.

The same goes with a microphone. Laptops have built-in microphones; you don't have to do anything to set them up. Desktops usually do not; you must add an external USB microphone. When you plug it in, Windows should recognize it automatically.

To take a photo with your webcam, use the Camera app. Follow these steps:

1. **Click the Start button, type** Camera, **and then click the Camera app in the search results.**

   If your camera is working you should see yourself — or whatever the camera is pointing at. Figure 15-1 points out some of the buttons that the rest of these steps refer to.

Take photo

Video

FIGURE 15-1

2. **Click the Take Photo button.**

   A snapshot appears as a small thumbnail image in the lower-right corner of the app window.

3. **Click the thumbnail image of the photo you just took.**

   It appears at a large size, so you can examine it.

4. **Do one of the following (see Figure 15-2):**

- To keep it and return to the Camera app's main interface, click the Back arrow (the left-pointing arrow) in the upper-left corner of the app window. The image exists in your Pictures ⇨ Camera Roll folder. You do not have to perform a separate action to save it.

- To delete it, click the Delete icon (the trash can) above the image.

- To share the image with someone, click the Share icon in the upper-right corner of the app window and use the controls that appear to share the image. For example, in Figure 15-2, I could choose Outlook Desktop to send someone an email using Outlook with this file as an attachment.

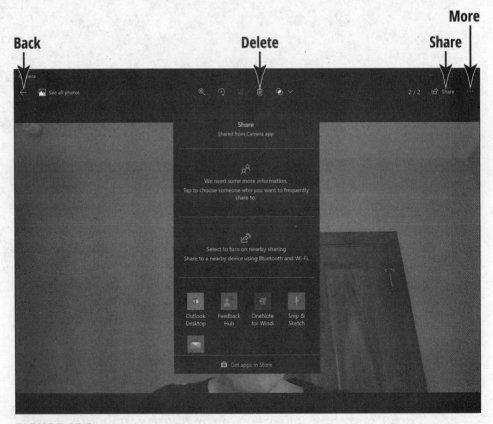

FIGURE 15-2

- Click the More icon (. . .) to open a menu and then choose Copy to copy the picture to the Windows Clipboard; you can then paste it in any app to use it there.

- Click See All Photos to open this picture in the Photos app and browse other photos at the same time. (More about the Photos app later.)

Notice in Figure 15-1 that the Take Photo button is larger than the Video button. That means you're in photo mode. Click the Video button to enter Video mode, where the Video button is the larger of the two.

The Camera app enables you to create and save videos of yourself (or whoever the webcam is pointed at) that include both sound and moving images. To do that, follow these steps:

1. **Click the Video button to enter Video mode, if needed.**

2. **Position the camera as needed.**

If it's an external camera, and you want to film something other than yourself, point the camera appropriately.

3. **Click the Video button to start the recording.**

A big, red, square Stop Taking Video button appears to the right of the image. See Figure 15-3.

4. **When you want to stop the recording, click the Stop Taking Video button.**

5. **If you want to preview the video before you save it, do the following:**

a. *Click the image thumbnail in the lower-right corner of the app window to open the video in a preview mode.*

b. *Click the Play button (the right-pointing triangle) at the bottom of the video. Make sure you hear any audio that was included in the recording.*

Stop taking video

FIGURE 15-3

6. **Do one of the following:**

- *To keep it and return to the Camera app's main interface, click the Back arrow (the left-pointing arrow) in the upper-left corner of the app window. The image exists in your Picture ⇨ Camera Roll folder. You do not have to perform a separate action to save it.*

- *To delete it, click the Delete icon (the trash can) above the image.*

- *To share the video with someone, click the Share icon in the upper-right corner of the app window and use the controls that appear to share the image.*

- *Click the More icon (. . .) to open a menu and then choose Open Folder to open File Explorer and display the folder where the video is stored (Pictures ⇨ Camera Roll).*

- *Click See All Photos to open this video in the Photos app and browse other photos at the same time.*

If you recorded a video that was supposed to include audio but you don't hear anything during playback, here are some things to check:

- Is the volume muted? Check the Volume icon in the notification area in the lower-right corner of the Windows desktop.

- If you have a desktop PC, are your external speakers connected to the PC? Do the speakers have power, and is the volume control on the speaker turned up?

- If you have a desktop PC, do you have an external microphone connected to the PC? Desktops don't usually have built-in microphones.

If all that checks out, run through the Microphone Setup utility in Windows. Click Start, type **Microphone**, and then click Microphone Setup in the search results. You can also check out the microphone settings in the Settings app: Settings ⇨ Privacy & security ⇨ Microphone. It is possible that someone disabled microphone access for privacy reasons.

Make Audio Recordings with Voice Recorder

The Voice Recorder app is a very simple utility that comes with Windows for making audio recordings using a microphone. You can use it to record audio for any purpose, like sending an audio birthday greeting to a friend via email or making a quick recording of some friends who are singing at a party. Nowadays, most people use their smartphones for this informal kind of recording, but it's good to know that your PC will do it also.

To make an audio recording, follow these steps:

1. **Click Start, type** Voice, **and then click Voice Recorder in the search results.**

The first time you use Voice Recorder (or if you don't have any saved recordings), the interface that appears is very simple, as in Figure 15-4. *Very* simple! As in, just a Record button in the center of the window.

FIGURE 15-4

2. **When you are ready to start the recording, click the Record button, and then start speaking (or doing whatever it is you want to record).**

3. **Click the Stop Recording button (the big square in the center).**

   You can also pause the recording by clicking the Pause Recording button, and then click Pause Recording again to restart it.

   After you've stopped the recording, the app's interface changes to offer some additional controls. See Figure 15-5. If your window doesn't look like Figure 15-5, try widening the window.

4. **Click Play to play back your recording.**

TIP

You can click the Add a Marker button (the flag) to mark a particular location so you can find it again easily later.

You can drag the slider under the Play button to jump to a certain spot in the playback.

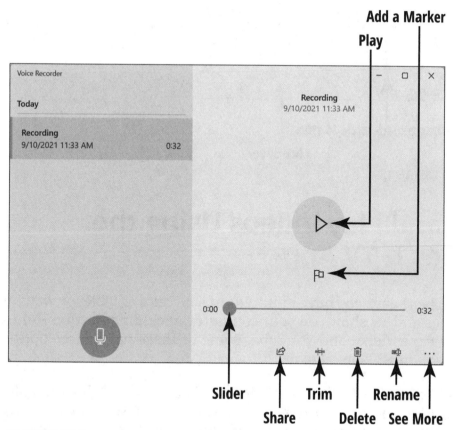

FIGURE 15-5

5. Do one of the following:

- *Click Share to share the recording with others.*

- *Click Trim to trim the recording to eliminate unwanted parts on the beginning and/or end. When you click Trim, two black circles appear on the timeline, as shown in Figure 15-6. Drag them to set the start and end points, and then click the Save icon to save the trimmed version. On the menu that appears, click either Update original or Save a copy.*

- *Click Delete to delete the recording.*

- *Click Rename to rename the recording.*

- *Click See More for a menu of other settings. For example, if your microphone isn't working properly, you can choose Microphone Settings to make adjustments.*

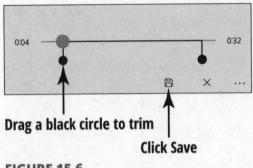

Drag a black circle to trim

Click Save

FIGURE 15-6

Find and Play Videos Using the Movies & TV App

These days you have hundreds of sources available for watching movies and TV shows on your computer, including Netflix, Hulu, Amazon Prime Video, and YouTube. Some of these have a free option; others you must subscribe to.

The Movies & TV app in Windows is only one of the available options, but it's an easy one since it's preinstalled in Windows and ready to go. This app works hand-in-hand with the Microsoft Store app, where you can buy or rent videos. The Store app handles the purchase, and then the Movies & TV app handles the playback. You can actually browse movies and TV shows in either of the two places — Windows will switch between the two apps as needed automatically. So that being the case, we'll start out in the Movies & TV app in the following steps.

To rent a movie and play it on your PC, follow these steps:

1. **Click Start, type** Movies, **and then click Movies & TV in the search results.**

2. **Click the Explore tab.**

Here you will find lots of content for sale or rent, mirroring the offerings you find in the Microsoft Store app.

3. **Click the title you want to buy or rent.**

The Microsoft Store app opens, showing you the purchase options for that title. See Figure 15-7.

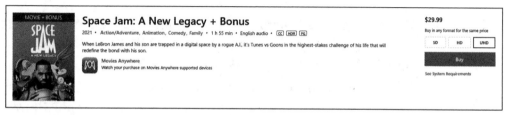

FIGURE 15-7

4. **Click the purchase option you want (for example, Rent).**

   A box appears asking how you want it delivered. See Figure 15-8. Your choices are to download it or stream it. Downloading it makes it available on your local PC so you can watch it even when you have no Internet connection (like on a plane). Streaming avoids taking up that space on your local hard disk, but requires an Internet connection.

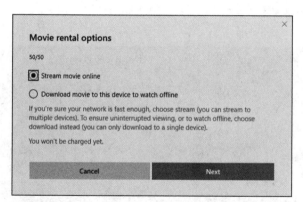

FIGURE 15-8

5. **Make your choice of delivery method and click Next.**

6. **If prompted, enter your security PIN.**

7. **Follow the prompts to complete the rental or purchase.**

   The steps will vary depending on what payment method(s) you have set up — or whether you still need to walk through the process of setting up a payment method.

8. **When the purchase is complete, return to the Movies & TV app and click the Purchased tab to see your purchases.**

9. **Click a purchased title and then click Play to begin watching it.**

10. Use tools at the bottom of the screen (see Figure 15-9) to do the following (if they disappear during playback, just move your mouse or tap the screen to display them again):

- *Adjust the volume* of any soundtrack by clicking the Show Volume Menu button to open the volume slider and dragging the slider left (to make it softer) or right (to make it louder). Click the megaphone-shaped volume icon to the left of the slider to mute the sound (and click it again to turn the sound back on).

- *Pause the playback* by clicking the Pause button in the center of the toolbar.

- *Enable captions* by clicking the Show Menu to Subtitles and Audio button and then clicking the language you prefer.

- *Change the display size of the movie.* Display the movie in a smaller window by clicking the Play in Mini View button.

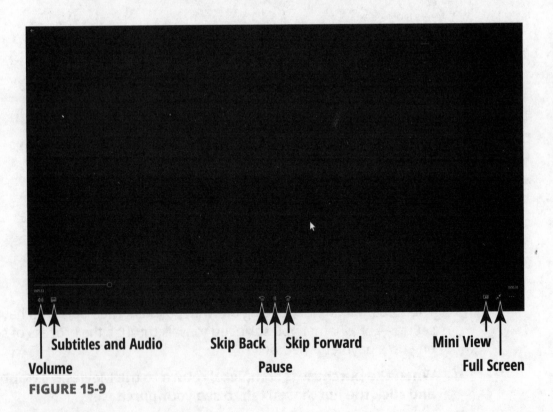

FIGURE 15-9

- *View the movie on full screen* by clicking the Full Screen button. When the movie is displayed in full screen, the Full Screen button becomes an Exit Full Screen button. Click this button to view the movie in a smaller window.

- *View the movie on a Bluetooth-enabled TV* by clicking the Show More Options button and then clicking Cast to Device.

11. **Click the Close button to close the Movies & TV app.**

TIP

You can also use the Movies & TV app to play your own video content. Your own content from your Video folder (C:\Users*username*\ Videos) will be on the Personal tab. The playback controls are almost identical to those discussed for playing movies.

STREAMING OR DOWNLOAD?

Today, many people prefer to stream video rather than download it. Playing a video on the YouTube website (a site where people can post their own videos) is one example of streaming, but you can also stream from sites such as Amazon or Netflix (sites that typically require a purchase or subscription to view movies or TV shows), just as you can from the Microsoft Store.

The advantage to streaming is that you don't have to store large video files on your device, especially important on a smartphone or tablet with smaller storage capabilities. The downside to streaming is, when you have a weak Internet connection, the playback can be glitchy, stopping and starting often.

If you have a strong Internet connection and want to try streaming, go to a site such as Amazon.com and locate a movie or TV show, rent or buy it, and start viewing.

Transfer Photos and Videos from a Camera or Phone

Uploading photos and videos from a camera or phone to your computer is a very simple process, but it helps to understand what's involved. (Check your manual for details.) Here are some highlights:

» **Making the connection:** Uploading photos and videos from a digital camera or phone to a computer requires that you connect the device to a USB port on your computer using a USB cable that typically comes with the camera or phone. Power on the camera or change its setting to a playback mode as instructed by your user's manual. Some cameras are Wi-Fi-enabled, which eliminates the need for a USB cable when you are in range of a Wi-Fi network.

» **Installing software:** Digital cameras also typically come with software that makes uploading photos to your computer easy. Install the software and then follow the easy-to-use instructions to upload photos and videos. If you're missing such software, or if you just prefer File Manager because it's familiar, you can simply connect your camera to your computer and use File Explorer to locate the camera device on your computer and copy and paste photo or video files into a folder on your hard drive.

» **Sharing photos from your phone:** You can share photos via a messaging app or email, and then open them on your computer using the same app. I often email a photo to myself from my phone's Mail app as a simple way to transfer it.

» **Printing photos straight from the camera:** Digital cameras save photos onto a memory card, and many printers include a slot where you can insert the memory card from the camera and print directly from it without having to upload pictures first. Some cameras also connect directly to printers. However, if you want to keep a copy of the photo and clear up space in your camera's memory, you should upload the photos to your computer or an external storage medium such as a DVD or USB stick, even if you can print without uploading.

View and Edit Photos in the Photos App

The Photos app provides a simple way to view, edit, and organize your personal photo collections. To browse your photos, Start the Photos app from the Start menu. It may be pinned to the Start menu; if not, begin typing Photos and then click it in the search results.

The Photos app provides several different ways of viewing your photos, each one designed for a certain type of management. The tabs across the top of the window include Collection, Albums, People, and Folders. Click each one to get a sense of how it presents your photos. For example, People groups the images based on who is in the photos, based on tagging and facial recognition. Folders groups the images based on their storage locations. Figure 15-10 shows the People view of my photos, When I click a person, all the photos with that person in them appear as thumbnail images, as in Figure 15-11.

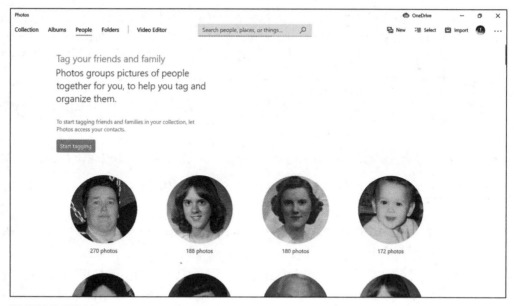

FIGURE 15-10

FIGURE 15-11

You can edit photos by performing such manipulations as rotating, adjusting brightness, and cropping. To edit a photo, double-click it to open it in an editing window. Then click the Edit & Create button to open a menu of options:

» **Edit:** Crop, rotate, flip, adjust brightness, or add filters, as shown in Figure 15-12.

» **Draw:** Use your mouse (or a touchscreen) to draw on the photo.

» **Add 3D effects:** Apply fun special effects like sparkles, snow, or confetti.

» **Add animated text:** Apply a text overlay that moves or fades in and out.

» **Create a video with music:** Opens the Video Editor, where you can build a slideshow with animation and music. Video Editor is covered in more detail later in this chapter.

» **Edit with Paint 3D:** Like the Draw tools but with more options, including adding stickers and 3D shapes.

FIGURE 15-12

Create a Video with the Video Editor

Maybe you've seen videos at retirement parties or memorial services that include pictures and videos of the person being honored, and wondered how someone put that together. There are lots of programs that you can use to do it. One of the easiest tools also happens to come with Windows: Video Editor.

A full tutorial on Video Editor could be its own chapter, but here's a very brief introduction that will get you started:

1. **Open Video Editor from the Start menu.**
2. **Click New Video Project.**
3. **Type a name for your video and click OK.**
4. **Click Add Title card. A blank slide with a colored background appears.**
5. **Click Text. A slide appears in editing mode, where you can put text.**
6. **Click a layout in the Layout section of the screen.**

7. In the Title box, type the text you want to appear.

8. (Optional) In the Animated text style area, click an animation style for the text. See Figure 15-13.

9. (Optional) To change the background color, click Background, and then click a color that you want to use for the title card.

10. Click Done.

FIGURE 15-13

11. In the Project Library section of the screen, click Add.

12. On the menu that appears, click where you want to pull content from: From this PC, From my collection, or From the web.

13. Browse for and select an item you want to include, such as picture of video clip. Click Open.

14. Drag the item from the Project Library area to the first empty slot in the Storyboard area.

15. Repeat steps 11-14 to add more photos or video clips. Figure 15-14 shows several items added to the timeline.

FIGURE 15-14

16. Click each item on the timeline and then click the Duration button and choose a duration for it to appear onscreen (if you don't want the default duration, which is 3 seconds).

17. Click the Play button (the right pointing triangle) to play a preview of the video.

18. Make any adjustments or additions as desired. There's a lot more you can do with this app than I have space to tell you about here.

19. Click Finish Video.

20. Click Export. The Save As dialog box opens.

21. Enter a name for the video file and choose a location for it. Then click Export.

22. When the video is finished, a preview of it plays in the Photos app window. Close that window when you are done watching it.

music

» **Introducing Windows Media Player**

» **Playing a digital music clip**

» **Creating a playlist**

» **Working with music CDs**

» **Exploring streaming music services**

» **Buying music in an online store**

Chapter **16**

Listening to Music on Your PC

M usic is the universal language, and your computer opens up many opportunities for appreciating it. Your computer makes it possible for you to listen to your favorite music, download music from the Internet, play audio DVDs, and organize your music by creating playlists.

You can set up your speakers and adjust volume, and then use the Windows Media Player app to play music and manage your music library.

Preparing to Listen to Digital Music

To listen to digital music, you will need either speakers or head-phones. Laptops have built-in speakers, so you're good-to-go. Desktop PCs require you to attach external speakers. Both laptops and desktops typically have a headphone jack that you can use to connect headphones; some PCs also have Bluetooth, a wireless network connection that enables you to connect wireless headphones.

Going the wired route is easy: just attach headphones or speakers to your computer by plugging them into the appropriate connection (often labeled with a little earphone or speaker symbol) on your tower, laptop, or all-in-one monitor. You're done! Move on to the next section.

To attach Bluetooth headphones, follow these steps:

1. **Open the Settings app.**

2. **Open the Bluetooth section of the Settings app by doing the following:**

   - *Windows 11:* Click Bluetooth & Devices.

   - *Windows 10:* Click Devices. The Bluetooth & Other Devices screen may appear automatically; if it doesn't, click Bluetooth & Other Devices.

3. **Click Add Device (Windows 11) or Add Bluetooth or Other Device (Windows 10).**

   The Add a Device dialog box opens. See Figure 16-1.

4. **Click Bluetooth.**

FIGURE 16-1

5. **Power on your Bluetooth headphones and place them in Pairing mode.**

There might be a pairing button on the headphones, or a slider that you move to a certain position. Consult the documentation that came with the headphones to find out how to do this. The dialog box shows a list of nearby Bluetooth devices, and this list should include your headphones.

6. **Click the headphones on the list of devices. See Figure 16-2.**

Windows pairs the headphones with the PC.

7. **Click Done.**

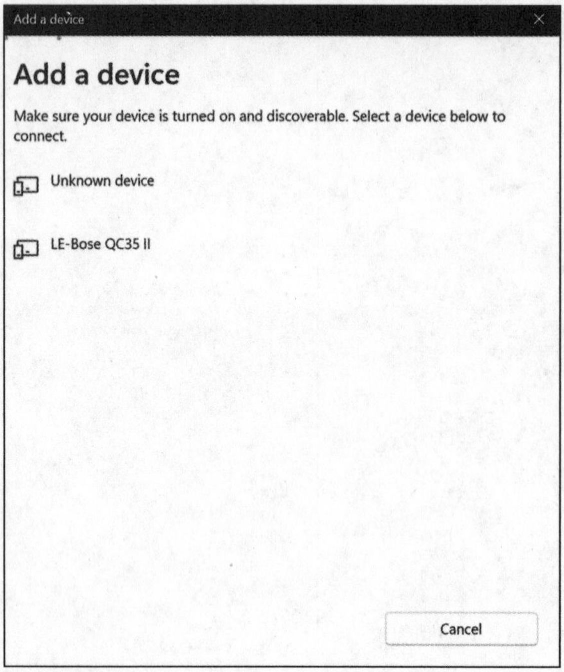

FIGURE 16-2

Next, you need to make sure the volume is turned up, both on the speakers (if applicable) and in Windows itself. On external speakers, there should be a volume control knob on one of them. Built-in speakers don't have their own volume controls; they take their cues from the operating system's settings.

To adjust the volume in Windows, click the Volume icon in the notification area (to the left of the clock in the bottom-right corner of the screen) and then drag the Volume slider to the right or left to make a change to the volume level. Figure 16-3 shows the Windows 11 version on the left and the Windows 10 version on the right.

TIP
By default, the volume is the same for the music and sounds played in every application. If you want a certain application to be louder or quieter than others, first make sure that application is open. Then right-click the Volume icon and choose Open Volume Mixer. In the panel that appears, you can drag sliders up and down for each open application that uses sound individually. Figures 16-4 and 16-5 show the Windows 11 and Windows 10 versions, respectively.

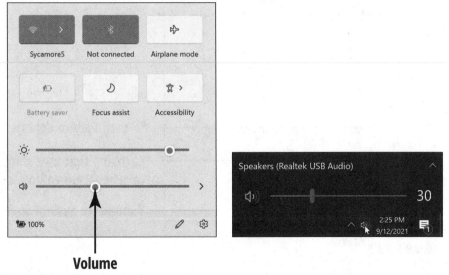

Volume

FIGURE 16-3

Volume

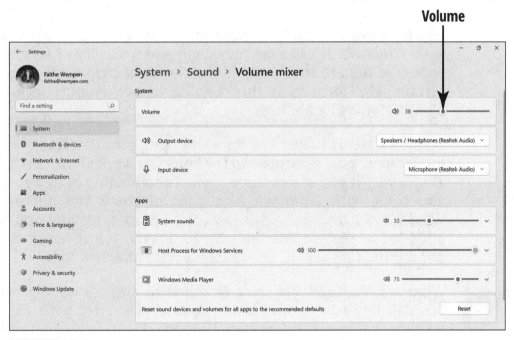

FIGURE 16-4

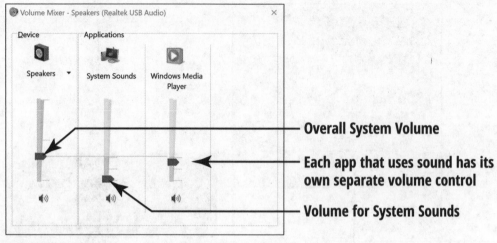

Overall System Volume

Each app that uses sound has its own separate volume control

Volume for System Sounds

FIGURE 16-5

Introducing Windows Media Player

Windows Media Player is a general-purpose media player that comes with Windows. It can play music clips in a variety of formats, show photos in a slide show, and given the right helper files, it can play certain video formats. In this chapter, we are mainly concerned with its music-playing abilities.

Windows Media Player is a "mature" program that has been around for decades, predating most of the apps you worked with in Chapter 15. That's why it duplicates some of the functionality of newer apps that also come with Windows, like video playback and photo viewing. Windows 8.1 included Groove Music, a music-playing app that was intended to replace Windows Media Player, but it never caught on, nor did the for-pay Groove Music service attached to it.

REMEMBER

Windows Media Player isn't attached to a music service where you can buy new music; it's strictly a player for your existing content that you may have acquired via downloads or by ripping (copying) CDs. See "Exploring Streaming Music Services" and "Buying Music in an Online Store" later in this chapter for more details about acquiring music online.

To start Windows Media Player, click Start, type **Media**, and then click Windows Media Player in the search results. Figure 16-6 shows the app and points out some of the key controls to know about:

» **Command bar:** Each of the buttons on this bar (Organize, Stream, Create Playlist) opens a menu of commands you can select.

There's also a menu bar for controlling the app, but it's hidden by default. To display it, click Organize, point to Layout, and click Show Menu Bar. We won't use it in this chapter.

» **Navigation pane:** Here's where you find quick shortcuts to key locations, including your playlists (covered later in the chapter), lists of your music sorted by artist, album, or genre, videos, and pictures.

» **Content pane:** A list of the content in whatever location or category you selected in the navigation pane.

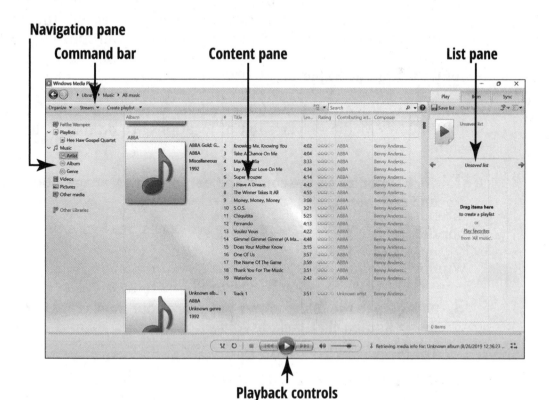

FIGURE 16-6

» **List pane:** A three-tabbed pane where you can drag-and-drop music clips to create lists:

- **Play:** Create a playlist.
- **Burn:** Create a list of music to burn to a writeable CD.
- **Sync:** Create a list of music to synchronize with a compatible portable music player.

» **Playback controls:** Basic controls for starting, stopping, skipping ahead, changing the volume, and so on.

Make Your Stored Music Available in Windows Media Player

If you store your music clips in your personal Music folder, they will automatically appear in Windows Media Player, because that location is automatically indexed. So, one way to make your music show up in Windows Media Player is to move or copy your music clips into that folder using File Explorer.

Another way is to add the folder where your music clips are stored to Windows Media Player's list of locations. To do that, follow these steps:

1. **Click Organize on the command bar, point to Manage Libraries, and click Music.**

   The Music Library Locations dialog box opens. See Figure 16-7.

2. **Click Add.**

   The Include Folder in Music dialog box opens.

3. **Navigate to the folder where you store your music clips and select it; then click Include Folder.**

   The folder is added to the list.

4. **Click OK to close the dialog box.**

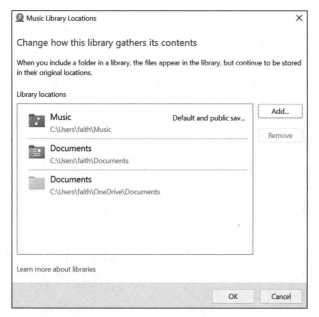

FIGURE 16-7

Play Music

Playing an individual music clip is very simple:

1. In the navigation pane, choose the general category of content you want to browse.

For example, under the Music heading, select how you want to view your clips: Artist, Album, or Genre. Browse until you find the track you want to play. Or, if you have playlists set up, click the desired playlist under the Playlists heading.

You can also search for a song, artist, or genre by entering what you want in the Search box above the content pane.

TIP

To go back to the previous screen, click the Back button (the left pointing arrow in the upper–left corner of the app window).

2. In the Content pane, double-click the track you want to play, or single-click it and then click the Play button in the playback controls at the bottom of the window.

If you displayed a list of songs (such as an album or a playlist), the whole list will play, item by item, unless you stop it early.

Figure 16-8 points out the individual buttons for controlling playback:

» **Play/Pause:** Plays the selected clip. If a clip is already playing, this button changes to Pause.

» **Previous:** Click to go back to the previous clip you played, or click and hold to rewind the current clip.

» **Next:** Click to go to the next clip (if you're playing a list of songs) or click and hold to fast-forward the current clip.

» **Volume:** Drag the slider to control the in-app volume. Unlike the general Volume control in Windows, this slider affects only Windows Media Player's volume.

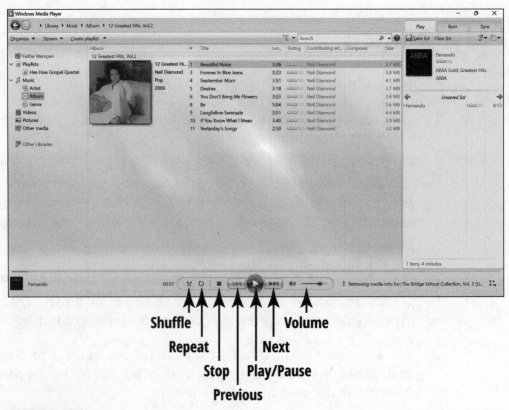

FIGURE 16-8

» **Stop:** Stops playback.

» **Repeat:** If you're playing a list of songs, turning on this feature makes it start over when it gets to the end.

» **Shuffle:** If you're playing a list of songs, turning on this feature plays them in random order.

Create a Playlist

A *playlist* is a saved set of music tracks you can create yourself — like building a personal music set, or "mix tape," to use the old-fashioned term for it.

You can create spontaneous playlists for one-time listening, or you can save your playlist for later replay.

In Windows Media Player, follow these steps to create a playlist:

1. **Make sure the List pane is visible, as in Figure 16-6.**

If it's not, click the Play tab in the upper-right corner of the app window to display it.

2. **Make sure the Play tab is selected in the List pane.**

3. **Using the controls you learned about in the previous section, browse for music clips you want to add to your playlist. When you find one, drag-and-drop it to the List pane.**

4. **Repeat step 3 as many times as needed to build your playlist.**

Figure 16-9 shows my Neil Diamond playlist, for example.

5. **Click the Play button to start the playback, or double-click one of the tracks in the playlist to start with that track.**

6. **(Optional) If you want the playlist to repeat, click the Repeat icon (refer to Figure 16-8).**

This feature is a toggle; if it is already on, clicking it turns it off.

7. **(Optional) If you want the playlist to play in random order, click the Shuffle icon (refer to Figure 16-8).**

This feature is a toggle; if it is already on, clicking it turns it off.

8. **(Optional) To save your playlist:**

a. *Click Save List.*

The current name (Untitled playlist) becomes highlighted, ready for you to edit it.

b. *Type the new name and press Enter.*

The new playlist's name appears in the navigation pane under the Playlists heading.

FIGURE 16-9

Rip a Music CD

Ripping may sound violent, but it's actually just a matter of convert-ing tracks from an audio CD to a digital format that can be stored on your PC. If your PC has an optical drive (that is, a drive that plays CDs and DVDs), you can rip your CDs to your hard drive so you don't have to physically insert the CD in the computer each time you want to enjoy that music.

TIP

No optical drive? You can buy an inexpensive external optical drive that connects to a USB port.

Follow these steps to rip an audio CD:

1. **Insert the audio CD in your PC's optical drive.**

2. **Locate the audio CD's name in the navigation pane below Other Media, and click it.**

Commands appear in the Command bar for ripping the CD.

3. **If you have a file format preference, click Rip Settings, point to Format, and choose the file format to use, as shown in Figure 16-10.**

TIP

MP3 is a good general-purpose format that does not have any copy protection and is easy to share. You will never run into any permis-sion issues with an MP3 file.

IS RIPPING LEGAL?

If you own the CD and you are ripping it for your own personal use, then yes, it is legal. It is illegal to share the tracks with others (although nobody is going to be knocking on your door to arrest you if you make a few mix CDs to share with friends and family), and it's illegal to make ripped tracks available online, such as on a bootleg file sharing service.

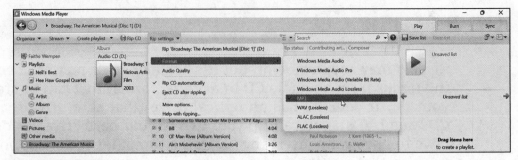

FIGURE 16-10

4. **Click Rip CD to begin the ripping process.**

It usually takes about 15 to 30 seconds per song. If you need to stop the rip before it completes, you can click Stop rip.

5. **Wait for the rip to complete. You will see the progress noted next to each file as it is ripped.**

You probably have a decent-sized collection of CDs you've accumulated over the years, and it may take you awhile to rip them all to your PC, but after you do, you'll enjoy the convenience of having all your music available, all at once.

Burn a Music CD

Burning is the opposite of ripping. Whereas ripping takes music from a CD and puts it on your hard drive, burning takes music from your hard drive and puts it on a CD.

To burn a music CD, follow these steps:

1. **Place a blank, writeable CD in your optical drive.**

It could be either a CD-R or CD-RW disc. CD-R are a bit more durable and long-lasting, but can't be rewritten later; whatever you write to them initially is what you are stuck with.

2. **Make your burn list by dragging tracks from the Content pane to the Burn tab of the List pane.**

   At the top of the burn list is an estimate of the remaining space on the disc. See Figure 16-11.

Remaining space on disc

FIGURE 16-11

3. **Click Start burn. Then wait for the burn to complete.**

   You'll know it's done because your optical drive will eject the disc.

4. **Write the disc's title on the non-shiny side of the disc with a Sharpie or other very soft marker.**

Acquire New Music

Want more music? There are two ways to go about that. You can buy music from an online store and download it, or you can subscribe to a streaming music service, paying a monthly fee to play any music you want to hear.

When you buy and download music, you own that track (or that album) forever. It exists on your hard disk, and you can play it whenever you like, whether connected to the Internet or not. This gives you the most control over your music collection, but it can also be expensive, since you must buy each album you want separately, usually at $15 or more per album. You can buy music downloads from Amazon and iTunes and from many private record labels and services. For example, some independent artists have their own websites where you can buy their tracks.

When you subscribe to a streaming service, you are renting the music, rather than buying it. It's available only when you are connected to the Internet, and only as long as you maintain your subscription. However, on the plus side, one subscription gives you access to as many as 75 million music tracks — probably just about any artist and album you can think of. For about $15 a month or so, you can play any music you want — even stuff you don't own and have no intention of buying. Some popular streaming services include iTunes, Amazon Music Unlimited, Spotify, and Pandora. Each of those except iTunes offers an ad-supported or content-limited free version.

Which one should you pick? For downloads, I like Amazon. They download in MP3 format, which is the format I prefer, and I don't have to go through the process of trying to convert the files to MP3 later. For streaming, I like Spotify, because of the deep catalog, the automatically created playlists that take my tastes into account, and the availability of a family membership. That said, these are just my preferences, and you may have great reasons for choosing others.

5

Windows Toolkit

IN THIS PART . . .

Working with home networks

Protecting and securing Windows

Maintaining and troubleshooting Windows

» Deciding what you want to share over a network

» Using Bluetooth

» Using a cell phone as an Internet hotspot

Chapter **17**

Working with Networks

Acomputer network enables you to share information and devices, such as a printer, among computers. If you share a single Internet connection with multiple computers in your home, you already have a home network. If not, you can easily build a network using common peripherals that you can buy at just about any store that has an electronics department.

This chapter explores several options for getting connected to other devices and sharing information using a network. You learn how to set up a wired or wireless network, how to configure your Windows PCs to share files on it, and how to work with wireless peripherals like Bluetooth devices. You also learn how to use a smartphone's Internet connection for a makeshift Wi-Fi network, which is useful if your main Internet connection is down and you urgently need to get online.

Plan and Set Up a Home Network

A home network can be pretty simple. Each computer (or other device) has a built-in network adapter that makes a connection to some sort of central gathering point, such as a router. Most home networks today are wireless — in other words, they use radio waves rather than cables. However, most wireless routers also have ports for connecting cables, so they can accommodate both wired and wireless connections.

If you have a broadband modem from your ISP that combines the functions of a modem and a wireless router, then you are all set as far as hardware goes. Ditto if you already have a wireless router that connects to your broadband modem. That's just two different ways of achieving the same thing: bringing in an Internet connection and then sharing it with all the computers in your household.

Look at each of your computers and figure out how it will (or already does) connect to the router: via Ethernet cable or via wireless (Wi-Fi) connection. Nearly every computer comes with a network adapter that offers at least one of those connection methods. If you see an Ethernet port on a computer, it accepts an Ethernet cable connection. Just plug one end of an Ethernet cable into the PC and the other end into a port on the back of the router (see Figure 17-1).

FIGURE 17-1

WARNING

If there's one port on the router that is a different color or offset from the others, don't use that one; it has a special function. That's the port that you connect your broadband modem to.

If there's no Ethernet port, then it's probably a laptop, and it probably has a wireless network adapter. To find out for sure, follow these steps:

1. **Right-click the Start button and click Device Manager.**

2. **Click Network Adapters to open a list of network adapters on your PC.**

3. **Look for one where the name includes the word *wireless*, or *802.11*, or *n/a/ac*.**

All these things indicate a Wi-Fi network adapter suitable for use on a home network. Figure 17-2 has one; can you find it there?

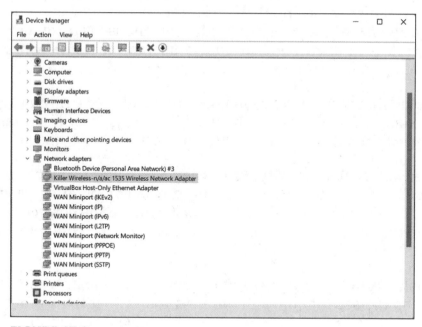

FIGURE 17-2

4. **While you're here, look for one where the name includes Bluetooth. If there is one, that information will come in handy later in the chapter.**

5. **Close Device Manager.**

6. **Turn back to Chapter 8 and work through the section "Set Up a Wi-Fi Internet Connection" to get your wireless connection up-and-running on your network.**

   The network shares the Internet connection, so by setting up the Internet connection, you are also setting up the network connection.

Enabling Wireless Router Security

Wireless routers are great because it's easy for any computer with a wireless network adapter to connect to them and participate in the network. Ironically, that's also the reason they are a dangerous security hazard. Anyone driving by your house can connect to your network, and anything that you've shared on the network — like private files — becomes fair game.

To plug that security hole, you need to enable security on your wireless router. It isn't enabled by default, so this is an important extra step you should take after getting the basic network up and running.

When you enable security, you assign a password (sometimes called a **key**) to the router. Then, the first time a computer or other device tries to connect to the router, it must provide the correct password. From that point on, the password is saved on the computer, and you don't have to enter it again.

I wish I could give you specific step-by-step instructions for setting up the security on your wireless router, but unfortunately each router's interface is different. I can only give you the general outline of what to do. Consult the documentation for your specific make and model of router for exact steps.

The general process goes like this:

1. **Open a browser window and enter the local network address for the router.**

   The router documentation should tell you what the default network address is for the router. You can also try some common addresses, like http://192.168.0.1 or http://192.168.1.1.

2. **If you see a prompt for a user name or password, consult the router's documentation to find out the defaults, and enter them. There might also be a sticker on the router that tells the defaults.**

3. **Poke around in the interface that appears to find where you can enter a password to secure the router.**

   Figure 17-3 shows mine, but yours may look very different. My router enables me to set up connections on two different frequencies (2.4GHz and 5Ghz), so that's why there are two separate sets of controls for the network name and the password.

4. **Look for a place where you can change the Admin password (the password you had to type in step 2). Change it to something you will remember.**

   If you leave it set to the default, anyone trying to hack into your network can easily guess it, because most router companies use generic, easy-to-guess defaults.

   On my router, the place where you can change the password is in a section called Management ⇨ Admin, but yours may be different.

   If you think you might forget the Admin password, write it on a sticky note and tape it to the bottom of the router.

TIP
5. **Save your changes (look for a Save button) and close the browser window.**

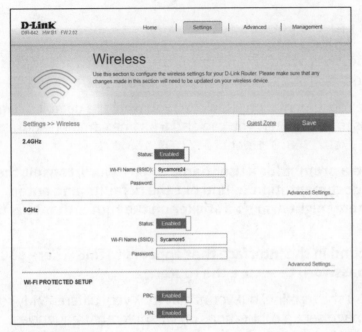

FIGURE 17-3

Set Up File Sharing on Your PC

File sharing is one of the reasons why people like home networks. However, just being connected to the network doesn't automatically share any files. You must decide what to share. That's a two-step process. First, you adjust the overall sharing settings as desired (and as explained here). Second, you choose individual folders to share (covered in the next section).

First, make sure that file sharing is enabled and your sharing settings are as you want them. To do that, follow these steps:

1. **Click Start, type** Control, **and click Control Panel in the search results.**

2. **In the Control Panel, click Network and Internet, and then click Network and Sharing Center.**

3. **In the navigation pane on the left, click Change Advanced Sharing Settings.**

4. **Click Private to expand the options under that heading.**

5. **Make sure that Turn on File and Printer Sharing is enabled, as in Figure 17-4.**

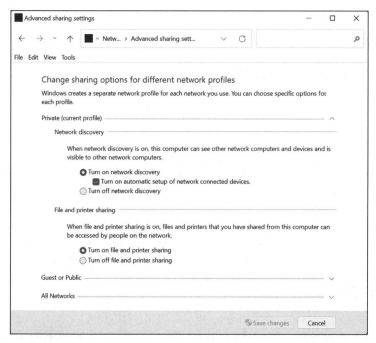

FIGURE 17-4

6. **Click All Networks to expand the options under that heading.**

7. **Turn Public Folder Sharing on or off as desired. See Figure 17-5.**

When this option is on, you can put files that you want everyone to be able to see and edit in the Public folders (C:\Users\Public). If you don't need a lot of privacy and you want minimal fuss when it comes to setting up sharing permissions, turn this on and then don't bother with any other sharing.

8. **Turn Password Protected Sharing on or off as desired.**

When this option is disabled, anyone on your local network can access your shared items; they need not have an account on your PC. This is a good option if you trust everyone on your network. The steps in the next section assume you have disabled this option.

When this option is enabled, only people who have a user account and password on this computer can connect to it remotely (via the network) to access the shared items. It provides for better control over who accesses your shared files and folders, but it requires you to set up user accounts on your PC for everyone who will be sharing your items. This is a good option if there are sometimes unknown guests on your network.

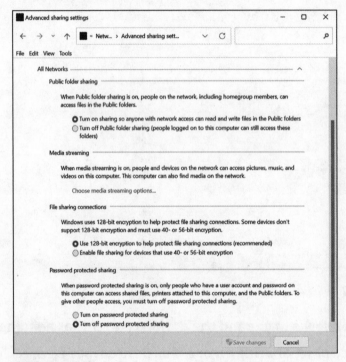

FIGURE 17-5

TIP

If you run into problems accessing shared resources between PCs, these sharing settings are the first place to look for answers. One troubleshooting technique is to set all these sharing settings to the most lenient settings temporarily. If the sharing inability clears up, you know it was one of the settings causing it.

Choose What Folders to Share

Now your PC is set up for sharing, but you haven't actually shared anything yet. (Well, if you turned on public sharing, you have shared the Public folder, but that's all.)

To share files with other people on your network, you put them in folders and then share the folders; you don't share individual files. So first off, identify a folder that you want to share. Then do the following:

1. Open File Explorer and select the local folder you want to share on your network.

By *local* I mean a folder on your hard drive (C:), not a OneDrive folder. You learned in Chapter 12 how to share a OneDrive folder.

2. Right-click the folder and click Properties.

The Properties dialog box for that folder opens.

3. Click the Sharing tab, and then click Advanced Sharing.

The Advanced Sharing dialog box opens.

4. Click to mark the Share This Folder check box. See Figure 17-6.

The folder is now shared with whoever has permission to access your network-shared items.

The preceding steps grant read-only access to a group called Everyone. Everyone is a built-in system-defined group in Windows that includes all users.

If you want the Everyone group to have write access, so people can change the content in that folder, do the following. Otherwise skip to step 9.

5. Click Permissions.

The Permissions dialog box opens for that folder. See Figure 17-7. Notice that the Everyone group has permission to the folder, and that group's permissions are set to Read but not to anything else.

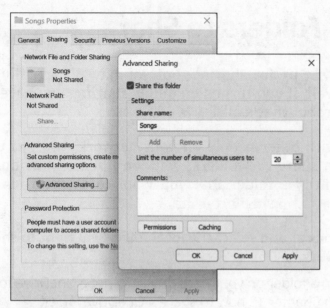

FIGURE 17-6

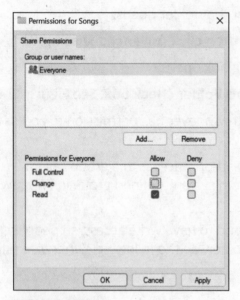

FIGURE 17-7

6. **In the Allow column, click to place a checkmark next to Change.**

This gives others the permission to change the content.

7. **(Optional) In the Allow column, click to place a checkmark next to Full Control.**

This gives others full permission not only to read and write but also to delete.

8. **Click OK to close the Permissions dialog box.**

9. **Click OK to close the Advanced Sharing dialog box.**

10. **Click Close to close the folder's Properties box.**

Browsing other people's shared folders is super easy. Just open File Explorer, click Network in the navigation bar on the left, and then double-click the PC containing the folder you want to share. You might be prompted for credentials if the option is enabled that requires you to have an account on that PC.

Figure 17-8 shows some shared folders on a PC named XPS. Notice that they have a green bar under the icons, indicating a network location.

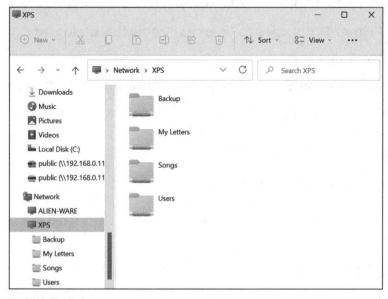

FIGURE 17-8

Share a Local Printer

A **local printer** is defined as one that is directly connected to one particular computer — in other words, the printer lacks its own network interface card that would make it directly accessible on the network.

It's better to have a printer with its own network interface, because then all PCs have access to it all the time. When a certain PC shares a local printer, others have access to it only when that PC is up and running. But if you don't have that option available, sharing a printer is the next-best way to make a single printer available to all.

To share a printer, follow these steps:

1. **Open the Settings app.**

2. **Do one of the following:**

   - *Windows 11:* Browse to Bluetooth & Devices ⇨ Printers & Scanners.
   - *Windows 10:* Browse to Devices ⇨ Printers & Scanners.

3. **Do one of the following:**

   - *Windows 11:* Click the printer you want to share.
   - *Windows 10:* Click the printer you want to share and click Manage.

4. **Click Printer Properties to open its Properties dialog box.**

5. **Click the Sharing tab.**

6. **Click to mark the Share This Printer check box. See Figure 17-9.**

7. **(Optional) Change the name in the Share Name box, if desired. This is the name other people will see.**

8. **Click OK.**

   Now the printer will be available for other people to set up.

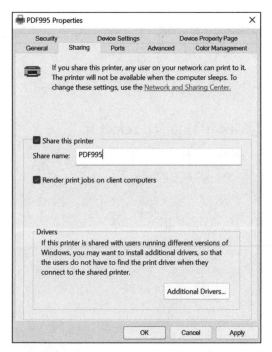

FIGURE 17-9

At this point you're only half done — you still must set up the print-er's driver on each PC that wants to share that printer. To do this, follow these steps:

1. Make sure the shared printer is powered on and connected to the PC that is sharing it.

2. Do one of the following:

- *Windows 11:* Browse to Bluetooth & Devices ⇨ Printers & Scanners.
- *Windows 10:* Browse to Devices ⇨ Printers & Scanners.

3. Click Add Device.

4. Do one of the following:

- *Windows 11:* Click Add Manually.
- *Windows 10:* Click The Printer That I Want Isn't Listed.

The Add Printer dialog box opens.

5. **Click Select a Shared Printer by Name.**

6. **Click Browse.**

7. **Double-click the computer to which the printer is directly connected.**

8. **Double-click the shared printer's icon to select it.**

   The Add Printer dialog box reappears. The printer's location and name are filled in, as shown in Figure 17-10.

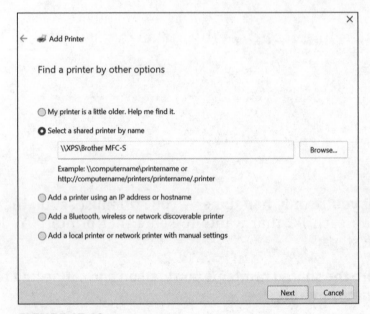

   ← 🖨 Add Printer   ✕

   Find a printer by other options

   ○ My printer is a little older. Help me find it.

   ● Select a shared printer by name

   | \\XPS\Brother MFC-S | Browse... |

   Example: \\computername\printername or
   http://computername/printers/printername/.printer

   ○ Add a printer using an IP address or hostname

   ○ Add a Bluetooth, wireless or network discoverable printer

   ○ Add a local printer or network printer with manual settings

                                          Next   Cancel

FIGURE 17-10

9. **Click Next to move to the next screen of the Add Printer dialog box.**

   A Printer Name text box appears. You can change the name of the printer here if you want. The name change will affect only your link to it, not the original.

10. **Click Next.**

11. **Click Finish.**

    The printer is now set up on the PC.

Connect Bluetooth Devices to Your PC

Bluetooth is a different kind of wireless networking. It's not for transferring files; it's for establishing connections between devices. For example, you might connect a Bluetooth printer, keyboard, mouse, or speakers. Bluetooth isn't compatible with Wi-Fi, but they can coexist side-by-side in the same environment. For example, you can have a Wi-Fi connection to your router (and the Internet) and a Bluetooth connection to your printer.

In Chapter 16, I explained how to connect Bluetooth headphones to your computer; the process of connecting other Bluetooth devices is similar. Follow these steps:

1. **Open the Settings app.**

2. **Open the Bluetooth section of the Settings app by doing the following:**

   - *Windows 11:* Click Bluetooth & Devices.

   - *Windows 10:* Click Devices. The Bluetooth & Other Devices screen may appear automatically; if it doesn't, click Bluetooth & Other Devices.

3. **Click Add Device (Windows 11) or Add Bluetooth or Other Device (Windows 10).**

   The Add a Device dialog box opens.

4. **Click Bluetooth.**

5. **Power on your Bluetooth device and place it in Pairing mode.**

   There might be a pairing button on the device, or a slider that you move to a certain position. Consult the documentation that came with the device to find out how to do this. The dialog box shows a list of nearby Bluetooth devices, and this list should include your device.

6. **Click the device you want to pair with the PC on the list of devices.**

Windows pairs the device with the PC.

7. **Click Done.**

Use Your Cell Phone as a Hotspot

When you need an Internet connection and one isn't available, it's easy to go into panic mode. We rely so much on the Internet for information, entertainment, and staying connected.

When the Internet is down at my house, I've been known to take my laptop and head out to a coffee shop or retail area where free Wi-Fi is offered. But if that's not practical, sometimes I set up my smartphone to function as a temporary Internet hotspot. A **hotspot** is a point you can connect to wirelessly to get network (Internet) access. Your wireless router at home is a hotspot, although we don't usually talk about home-based connections that way.

Most smartphones have a feature you can use to turn the phone into a portable wireless router temporarily, allowing wireless devices (like laptops and tablets) to use its Internet connection. This is sometimes called **tethering** because the computer is temporarily connected (tethered) to the phone.

WARNING

If you don't have unlimited data on your cell phone plan, or if your plan slows down your Internet speed after you've used a certain amount of data, be aware of the amount of bandwidth the connected computer is using while tethered to the phone. You might want to avoid things like watching videos, for example, until your regular Internet connection is restored.

Start by turning on the hotspot feature on your phone. The steps for doing this vary depending on the phone's operating system and version.

» **iPhone/iOS:** Settings ⇨ Personal Hotspot ⇨ Allow Others to Join. You can optionally set a Wi-Fi password so random strangers won't try to join.

» **Android:** Settings ⇨ Network & Internet ⇨ Hotspot & Tethering ⇨ Wi-Fi Hotspot. The page that appears enables you to set the hotspot name, set a password, and so on.

When the hotspot is enabled, it appears in the list of available Wi-Fi networks. See "Set Up a Wireless Internet Connection" in Chapter 8 to recall how to connect to one.

WARNING

Be aware of the drain on your phone's battery when tethering. Connect your phone to a power source when tethering, if possible, and turn off the hotspot when you're not using it.

~~Supplied or third-party security~~
software

» **Updating Windows**

» **Checking Windows security settings**

» **Changing your Microsoft account password**

» **Changing how you sign into Windows**

Chapter **18**

Protecting Windows

Your computer contains software and files that can be damaged in several different ways. One major source of damage is from malicious attacks that are delivered via the Internet.

Microsoft provides security features within Windows that help to keep your computer and information safe, whether you're at home or travelling with a laptop computer. In addition, there are software programs you can purchase or find for free online to monitor, block, and repair unwanted attacks on your computer.

In this chapter, I introduce you to the major concepts of computer security and cover Windows security features that enable you to do the following:

» Update Windows with the latest available security and performance fixes

» Check Windows security settings across a variety of utility categories, including anti-malware and firewall protections

» Change the password you use to sign into Windows

Choose Security Software

Every day you carry around a wallet full of cash and credit cards, and you take certain measures to protect its contents. Your computer also contains valuable items in the form of data, and it's just as important that you protect that information from thieves and damage.

Some people create damaging programs called **viruses** specifically designed to get into your computer's hard drive and destroy or scramble data. Companies might download **adware** on your computer, which causes pop-up ads to appear, slowing down your computer's performance. Spyware is another form of malicious software that you might download by clicking a link or opening a file attachment; **spyware** sits on your computer and tracks your activities, whether for use by a legitimate company in selling products to you or by a criminal element to steal your identity. Collectively all these different software-based threats are known as **malware**.

Microsoft Windows includes security features to protect your system and its valuable data. You can use Windows security tools such as Windows Defender to protect your computer from malware. Other security tools in Windows, such as the Windows Defender Firewall and the Credential Manager, protect your system in other ways. In this chapter, I focus on the security tools that come with Windows, but instead of relying solely on Windows security tools, you can use a third-party anti-malware program to protect your computer. There are several commercial products on the market, such as security suites by McAfee and Symantec, which charge a yearly subscription fee. The actual names of the security suites are a constantly moving target; it seems like both McAfee and Symantec change the names of their suites every year, and sometimes have multiple suites on the market simultaneously, so just look for the brand names. Figure 18-1 shows McAfee LiveSafe running a scan for viruses.

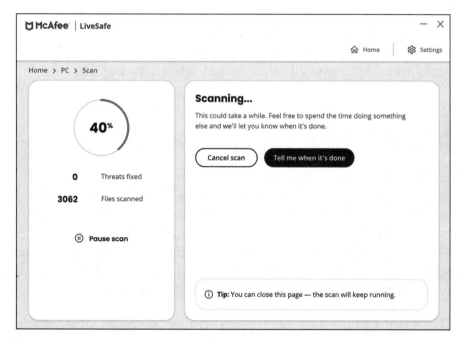

FIGURE 18-1

TIP

Don't discount the security tools in Windows just because they're free. They're just as effective on a basic level as any of the for-pay services. The main differences between Windows security tools and third-party tools are in two areas:

» **Customization:** Third-party security suites generally offer more customization options for the tools they include that more or less duplicate Windows security tools, like the firewall and the anti-malware software.

» **Scope of features:** One advantage of these third-party tools over the ones that come with Windows is that they often offer additional security tools that protect you in different ways. For example, some of them offer a browser add-in that helps you avoid sites with bad reputations, add-ins for Microsoft Outlook and other popular email clients that can scan for spam, and identity protection services and/or password management utilities.

Free security products are also available. Search the Internet for **best free antivirus programs** to locate one. Many of these programs are excellent, but generally you do get what you pay for.

Update Windows

When a new operating system version is released, it has been thoroughly tested — at least as far as the developers are able. However, when the product is in general use, the developers begin to get feedback about problems or security gaps that they couldn't anticipate. For that reason, companies such as Microsoft release updates to their software, both to fix those problems and deal with new threats to computers that appear after the software release.

Windows Update is a tool you can use to make sure your computer has the most up-to-date security measures in place. Today, most updates happen automatically, but there are some options you can adjust.

To view the available update options and change their settings, if desired, follow these steps:

1. **Open the Settings app and then:**

   Windows 11: Click Windows Update.

   Windows 10: Click Update & Security.

   Windows automatically checks for updates, and if any are available, it downloads and installs them. It actually does this automatically even if you don't open the app; it just does it on its own schedule. If you want it to check again, click the Check for Updates button. See Figure 18-2.

   If there's an Install Now button, click it to install the downloaded update. (If an update requires a restart, or if it's a major features update, Windows may delay installing it until you click Install Now.)

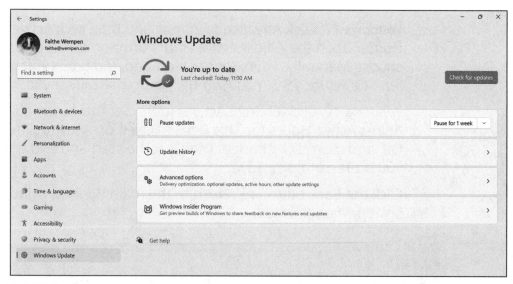

FIGURE 18-2

2. **(Optional) To temporarily pause updates:**

   Windows 11: Click the Pause for 1 Week button. You can also change the pause duration by opening the drop-down list associated with the button.

   Windows 10: Click Pause Updates for 7 Days. If you want to control the duration, click Advanced Options and then, under the Pause Updates heading, open the Pause Until drop-down list and select a date (which can be up to 35 days in the future). Click the Back button to return to the main Windows Update screen when finished.

REMEMBER

You cannot turn Windows Update off entirely; you can only pause it temporarily. You might want to pause getting updates if you are going to be traveling, for example, and relying on your cell phone's Internet connection.

3. **Windows Update installs updates at times when you are not likely to be using the computer — that is, outside of** active hours. **Windows sets active hours automatically by default. To make a manual change:**

Windows 11: Click Advanced Options and then click Active Hours. Open the Adjust Active Hours drop-down list and choose Manually. Then, set the start and end times you prefer. See Figure 18-3. Your setting is automatically saved.

Windows 10: Click Change Active Hours. Set the Automatically Adjust Active Hours for This Device Based on Activity slider to Off, and then click Change. Then, set the start and end times you prefer and click Save.

Click the Back button to return to the main Windows Update screen when finished.

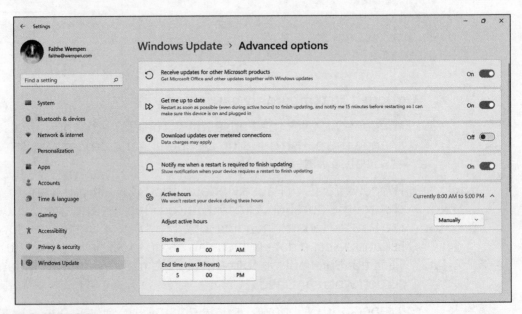

FIGURE 18-3

4. **(Optional) To choose whether to also receive updates automatically for Microsoft products, click Advanced Options and then set the Receive Updates for Other Microsoft Products slider to either On or Off.**

In Windows 10, the slider's name is slightly longer: Receive Updates for Other Microsoft Products When You Update Windows. It's the same other than that, though.

5. **Set any other options as desired.**

I won't go through them all in detail here, but you can find them in the Advanced Options section. Figure 18-3 shows the Advanced Options section for Windows 11, and Figure 18-4 shows it for Windows 10.

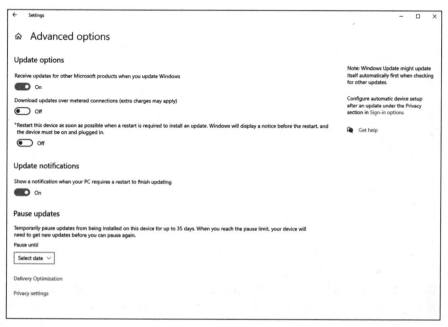

FIGURE 18-4

Check Windows Security Settings

In earlier Windows generations, checking security could get complicated because of the multiple utilities you had to look at. Today, though, a single app called Windows Security provides a single-pane dashboard where you can see all the various security settings at a glance. See Figure 18-5.

To run Windows Security, click Start, type Security, and then click Windows Security in the search results.

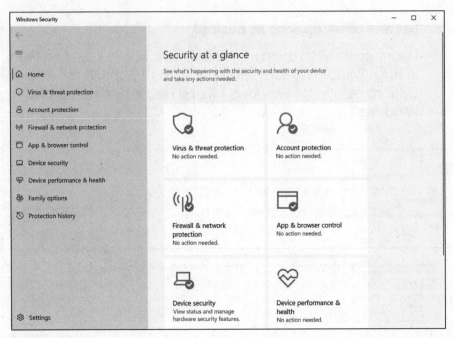

FIGURE 18-5

Here's a quick explanation of each of the categories you see in Figure 18-5. You can click each one to open a page where you can see its status and change some of its settings. This is optional reading, though, because as long as each one reports *No action needed*, you're good to go:

» **Virus & Threat Protection:** Here's where you check and configure Windows Defender, the anti-malware utility. It is enabled and runs automatically by default. You can do a quick scan, set scan options, and manage threat protection settings.

» **Account Protection:** Here you can see what Microsoft account you're signed in with, configure Windows Hello (a feature that lets you control how you sign in), and turn on Dynamic Lock, a feature that automatically locks your PC when you step away from it.

» **Firewall & Network Protection:** Here you can check to make sure the Windows Defender Firewall is enabled across all types of networks. If you are having a problem with the firewall wrongly blocking an app, you can set up an exception for it here.

- » **App & Browser Control:** Here you can view and control settings for blocking harmful content in your web browsing sessions, including reputation-based protection and exploit protection. When you select each of those, screens with more information about the concepts appear, so explore these on your own if you're curious.

- » **Device Security:** This screen reports on the security features built into your computer. Most of this is information-only.

- » **Device Performance & Health:** This screen reports on your computer's key functions that may affect your user experience, such as Windows Time Service (is it updating the time periodically?), Storage Capacity (are you running out of disk space?), Battery Life (is your battery holding a charge okay?), and Apps and Software (are any apps causing problems?)

- » **Family Options:** Here you can view and change parental controls that affect any user accounts set up as Child accounts. I explained this in Chapter 2 in the section "Manage Family Settings."

- » **Protection History:** Here you will see any actions that have been recently taken to help with security.

Change Your Microsoft Account Password

Most people use a Microsoft account to sign into Windows. That account might seem like it is local to your individual PC, but it is actually stored and managed on a Microsoft-owned database server on line. That's useful because your settings carry over between computers that way. For example, suppose you have two computers: one at your main residence and one at your vacation cottage. You create an account on both PCs using the same Microsoft account credentials, and any changes you make on one computer (like changing the desktop background or saving a file to OneDrive) are automatically picked up on the other computer.

If your password for this important account gets compromised in some way — maybe someone is snooping and sees you type it, or you tell it to someone in a moment of weakness and then regret doing so — you can change it. Here's how:

1. **Make sure you are signed in with the Microsoft account for which you want to change the password.**
2. **Open the Settings app and click Accounts.**
3. **Click Your Info.**
4. **At the bottom of the window, in the Related Settings section, click Accounts.**

   A web page opens where you can manage your Microsoft account.
5. **Click the Change Password link.**

   A prompt appears where you can change your password. See Figure 18-6.
6. **Fill in the prompts and click Save.**

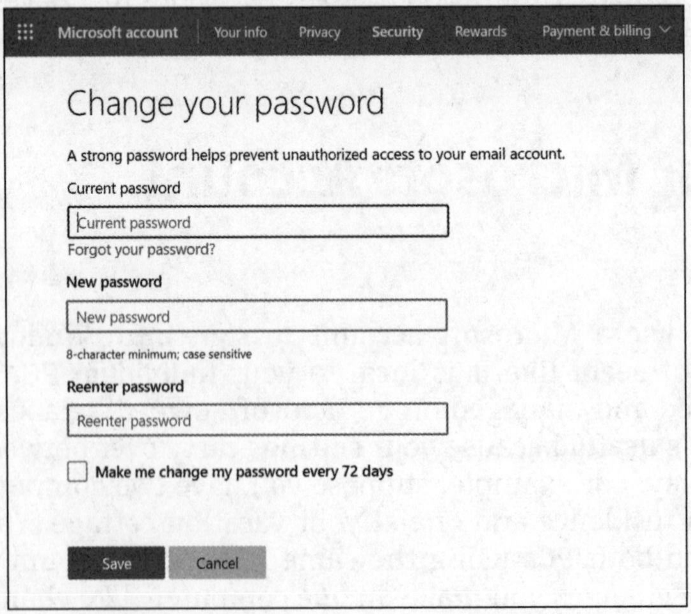

Microsoft account
Your info
Privacy
Security
Rewards
Payment & billing
Change your password
A strong password helps prevent unauthorized access to your email account.
Current password
Current password
Forgot your password?
New password
New password
8-character minimum; case sensitive
Reenter password
Reenter password
Make me change my password every 72 days
Save
Cancel

FIGURE 18-6

Change How You Sign Into Windows

If you have a long password (and you really should, for security reasons), it can be a pain to type it every time you sign into Windows, right? Fortunately Windows offers several other less onerous sign-in methods that you can configure on a per-PC basis. (In other words, if you use the same Microsoft account to sign into multiple machines, you can use a different sign-in method on each one if you like.)

The available sign-in methods will vary depending on your PC's hardware capabilities, but may include:

» **Facial recognition:** Your webcam reads the face of the person signing in, and if it matches up with a stored photo, you don't have to type a password.

» **Fingerprint recognition:** The fingerprint reader on your PC reads your fingerprint, and if it matches up with a stored fingerprint, you don't have to type a password.

» **PIN:** You can specify a PIN (a number) to type to sign in rather than your full password. This is usually a lot less typing!

» **Security key:** You can sign in with a physical device that acts as a security key, such as a USB fob.

» **Picture password:** You can store a photo that will appear on the sign-in screen; then you interact with the photo by clicking or dragging across various parts of it to sign in. It's like a password, but with movement rather than characters.

To access all these sign-in methods, pick the one you want to use, and configure it, follow these steps:

1. **Open the Settings app and click Accounts.**

2. **Click Sign-in Options.**

   A list of sign-in options appears. See Figure 18-7.

3. **Click the option you want to use.**

   For example, in Figure 18-7, I chose PIN.

4. **Use the controls that appear to configure that option.**

For example, in Figure 18-7, I can change my PIN or remove the option to sign in with a PIN.

5. **Scroll down to view the other sign-in options under Additional Settings and make any changes you like.**

For example, you can specify when Windows should require you to sign in again after being away, control whether restartable apps should restart when you sign back in, and use your sign-in information to allow Windows to restart after an update.

6. **Close the Settings app when you are finished.**

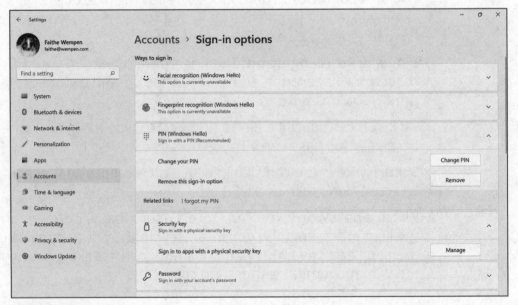

FIGURE 18-7

Chapter **19**

Maintaining Windows

All the wonderful hardware that you've spent your hard-earned money on doesn't mean a thing if the software driving it goes flooey. If any program causes your system to **crash** (meaning it freezes up and you must take drastic measures to revive it), you can try a variety of steps to fix it. You can also keep your system in good shape to help avoid those crashes.

This chapter explains various ways to recover and troubleshoot when you have problems with an app, or with Windows itself. You'll also learn how to improve system performance by freeing up disk space.

Shut Down an Unresponsive Application

When an application stops working, usually it's the result of a few different scenarios:

» **The app just closed unexpectedly, all by itself.** When you try to restart it, it might start normally, or it might not. If it doesn't, reboot your PC.

» **The app stopped responding to commands. A white film has appeared over its window to indicate it is locked up.** An error message appears asking whether you want to shut it down or wait. You can try waiting if you want — sometimes an app does bounce back after a minute or two — or you can shut it down right away and restart it. The only time I would wait it out would be if I had unsaved work in the application. If the application stops working again, reboot.

» **The app stopped responding to commands. It just sits there.** Other apps may start running sluggishly because the offending app is using up more than its share of CPU and memory resources.

It's that third scenario that I want to talk about here, because you actually have to do something to shut it down. Follow these steps:

1. **Press Ctrl+Shift+Esc.**

The Task Manager window opens. This window lists all running applications. It has two views: Fewer Details and More Details. If the list looks very plain and sparse like in Figure 19-1, you are in Fewer Details view. If you see tabs across the top of the window, you are in More Details view, as in Figure 19-2.

REMEMBER

The contrast between Figures 19-1 and 19-2 points out something important: not every running program always appears in Fewer Details view. In Figure 19-1, only two apps are listed as running, but in Figure 19-2 there are six.

TIP

If pressing Ctrl+Shift+Esc doesn't bring up the Task Manager, you're in bigger trouble than you thought. You may need to press and hold your computer's power button for five seconds (or slightly longer) to shut down.

FIGURE 19-1

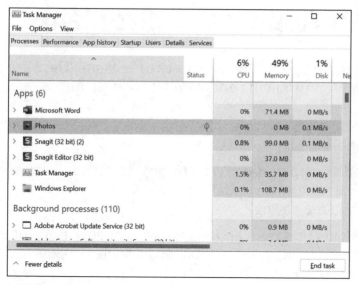

FIGURE 19-2

2. **Click the app you want to shut down to select it.**

If you don't see it, and you are in Fewer Details view, click More Details and look for it on the longer list that appears. You can sort this list alphabetically by name (if it's not already so) by clicking the top of the Name column.

3. **Click the End Task button.**

   The app shuts down.

   A dialog box may appear when an application shuts down asking if you want to report the problem to Microsoft. If you click Yes, information is sent to Microsoft to help it provide advice or fix the problem down the road.

4. **Close the Task Manager window.**

5. **Reboot your PC.**

   This is optional, but it's a good idea. It flushes out any leftover gunk that might still be in memory.

Troubleshoot Application Problems

Applications mess up for a wide variety of reasons. Here are some of the most common problems and how to solve them.

» **Reboot.** You will be amazed how often this solves a problem with an app.

» **Download an app update.** Visit the app maker's website and look around to see if an update is available that may fix the problem you are having.

» **Run Windows Update.** Windows usually updates itself automatically, but you can trigger it to look for available updates to speed the process. I cover this in Chapter 18.

» **Update your display adapter's driver.** This might not seem intuitive, but it often fixes problems with graphics-heavy applications such as full-featured shoot-em-up games. It works because often it's not the app's software that is faulty, but the connection between that software and the driver for a piece of hardware it uses. The most common culprit in such cases is the display adapter.

>> **Repair the app.** Sometimes you can repair an app using its setup utility without having to completely remove and reinstall it. See "Repair or Remove an App" later in this chapter.

>> **Remove and reinstall the app.** If repairing isn't possible or doesn't work, this is your next step. See "Repair or Remove an App" later in this chapter.

>> **Use Compatibility Mode.** Very old apps created for earlier Windows versions may not run well under modern Windows versions without some tweaking. See "Set an App to Run in Compatibility Mode" later in this chapter.

Repair or Remove an App

Sometimes a problem with an app can be fixed by repairing it (less drastic) or removing and reinstalling it (more drastic). You can do both from the Settings app:

1. **Open the Settings app and choose Apps ⇨ Apps and Features.**

2. **Find the app on the list, and click its More button (the three vertical dots at the far right).**

   A menu appears of valid commands for that app. Depending on the app, these may include Modify, Repair, Uninstall, Advanced Options, or some combination of all four.

3. **If a Repair command is available, choose it. Then walk through the prompts to repair the app.**

   If that doesn't work, go to the next step.

4. **If a Modify command is available, choose it. Then walk through the prompts.**

   During the modification process, there may be a prompt to repair the app, as shown in Figure 19-3.

   If that doesn't work, go to the next step.

5. **If an Advanced Options command is available, choose it. Then, browse through the available options for that app.**

   For example, you might be able to repair, reset, or uninstall it from this page.

6. **Click Uninstall. Then walk through the prompts to remove the app.**

7. **Reinstall the app using the same media you originally used for it (such as a download or an optical disc).**

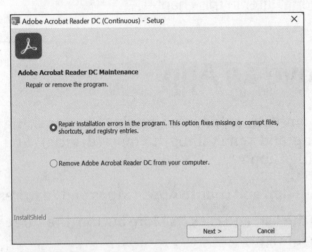

FIGURE 19-3

Set an App to Run in Compatibility Mode

Very old apps written when earlier Windows versions were the norm may have some programming quirks that rely on those earlier versions, and may not work well under Windows 10 or 11. If you really need to run one of these old programs (like a favorite game from way back in the day), you might be able to trick it into working by using Compatibility Mode. Compatibility Mode sets up conditions in Windows that makes the app think it is running in the earlier version.

To set up an app for Compatibility Mode, you need access to either the executable file that runs the app (that is, the program file, the one with a .com or .exe extension) or you need access to a shortcut to the file, such as a shortcut on the desktop. After you've located a suitable file for the app, follow these steps:

1. **Right-click the file.**

2. **If you're using Windows 11, click Show More Options. (Windows 10 users don't have to do anything here.)**

3. **Click Troubleshoot Compatibility.**

   The Program Compatibility Troubleshooter runs.

   You can use the compatibility troubleshooter in either of two ways: You can allow it to recommend settings (and then try those settings) and keep doing that until you find the settings that work. Alternatively, you can choose to troubleshoot based on the problems you are noticing. See Figure 19-4. Either way, work through the prompts that appear to guide you.

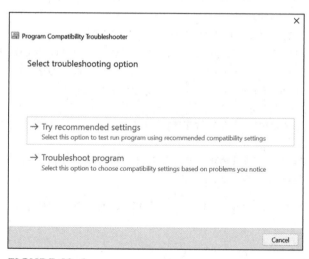

FIGURE 19-4

Restore Your System Files

Suppose that Windows and all your apps are working great one day — and then the next day something is gravely wrong. Windows seems sluggish or unstable, and your favorite apps take forever to load — or don't load at all.

Chances are good that something has changed your system files and the change is causing a problem. This can be anything from a faulty driver update for a piece of hardware to some malware that slipped into your PC via a shady website. Whatever the reason, you are wishing that you had your old system configuration files back, right?

Well, here's the thing. Old versions of those files probably *are* available. Windows has a feature called System Protection that backs up your system files automatically at certain times, such as before installing Windows updates. Depending on how your system is configured, it may create daily or weekly system restore points too, just on general principles. It also enables you to create your own backups of system files any time you like. Then, when you run into a problem, you can restore your system files to an earlier time — yes, you really can turn back time!

But first, let's think proactively. Let's say you are planning to install a new driver for your display adapter that you got on a third-party website — not the website for the manufacturer. It could be fine — or it could cause a problem that would be difficult to recover from. Your best bet would be to manually create a restore point before you install it. That way it will be easy to go back if needed.

Follow these steps to create a restore point:

1. **Open the Start menu and type** Restore, **and then click Create a Restore Point in the search results.**

   The System Properties dialog box opens to the System Protection tab.

2. **Click Create.**

   The System Protection dialog box opens.

3. **Type a descriptive name for the restore point, such as** Before Installing Driver. **See Figure 19-5.**

4. **Click Create.**

   The restore point is created. A confirmation box appears when it's finished.

5. **Click Close.**

6. **Click OK to close the System Properties dialog box.**

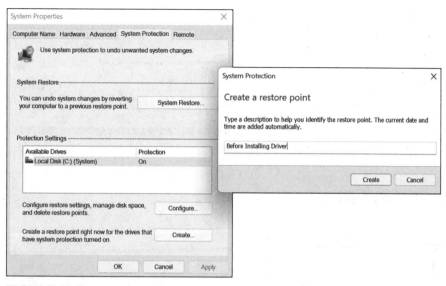

FIGURE 19-5

Now flash forward to the time when you need to use the restore point. Your driver installation has caused major stability problems with Windows, and you want to go back in time.

Follow these steps to restore system files to an earlier version:

1. **Open the Start menu and type** Restore, **and then click Create a Restore Point in the search results.**

The System Properties dialog box opens to the System Protection tab.

2. **Click System Restore.**

The System Restore dialog box opens.

3. **Click Next.**

4. **Click one of the saved restore points on the list that appears. See Figure 19-6.**

Choose the newest point that is older than the time when the problem began. If you're concerned about what changes will happen, click the Scan for Affected Programs button.

If you don't see any appropriate restore points, mark the Show More Restore Points check box to see older points.

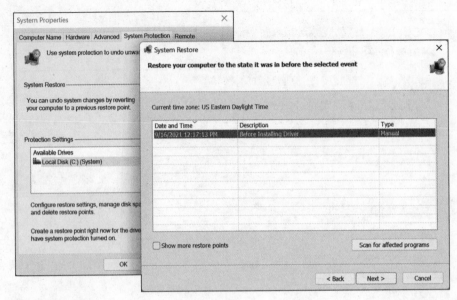

FIGURE 19-6

5. **Click Next.**

6. **Click Finish.**

7. **Click Yes to confirm.**

   The restore process begins. Wait for it to complete. You can't stop it after it starts.

8. **When the PC restarts, a confirmation box appears. Click Close to close it.**

Use Windows Troubleshooter Utilities

If System Restore didn't resolve your problem, the next step is to use Windows' troubleshooting utilities. There are separate troubleshooters for each of a dozen or so different kinds of problems.

1. **Open the Settings app and navigate to System ⇨ Recovery.**

2. **Click Fix Problems Without Resetting Your PC.**

3. **Click Other Troubleshooters.**

   A list of troubleshooting utilities appears.

4. **Click Run for the troubleshooter that best fits the problem you are having.**

   Figure 19-7 shows an example.

5. **Work through the steps that appear. The steps are different for each troubleshooter.**

FIGURE 19-7

Reset Your PC: The Last Resort

If you've tried System Restore and it didn't help your Windows woes, the last self-help trick in your bag is to reset the PC. Resetting the PC is no small affair, and not to be lightly undertaken. It resets Windows back to its original version, and wipes out all your installed apps and personal data files. You essentially start from scratch with a brand-new copy of Windows, like on the day you unboxed the computer.

Before you think about resetting, you should, of course, make sure that you have backed up as much of your personal data as possible. Because I save all my important data files to OneDrive, I don't have to worry about this because they're automatically mirrored from the local drive to OneDrive. But if you save your files on your local hard disk, you will want to back them up.

You should also consider how you plan to reinstall your applications after the reset. If you bought them on DVD, you should make sure you have the discs. If you bought them online, make sure you have any URLs, access keys, or registration numbers needed to get them back.

After all that planning and preparation, you're ready to do the reset. Take a deep breath and let's get started.

TIP

If your PC is so far gone that it won't even start normally, it starts in Windows Recovery Environment instead. There you'll find a menu of options, and one of the options is to reset your PC. So that's an alternate way of getting started with the reset process.

1. **Open the Settings app and choose System ⇨ Recovery.**

2. **Click Reset this PC.**

You're asked whether you want to keep your personal files or remove everything. See Figure 19-8.

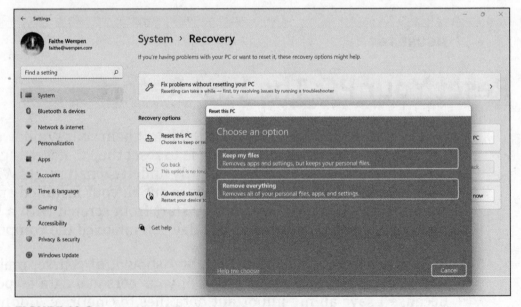

FIGURE 19-8

3. **Click the option you prefer.**

   Next you're asked whether you want to reinstall Windows via a cloud download or a local reinstall.

   A cloud download is the best way to get the latest version of Windows, but it takes longer because of the download. If you don't have unlimited Internet data transfer privileges with your ISP, such a large download could also potentially incur extra charges or slow down your connection speed as a penalty. (This is mostly the case with cellular and satellite Internet service.)

   A local download requires you to have an optical disc or some other form of local media from which to pull the Windows installation files.

4. **Click the option you prefer.**

   You see an Additional Settings window next, summarizing the choices you've made.

5. **If everything looks all right, click Next. Otherwise, click Change Settings to make changes.**

6. **Follow the prompts that appear to complete the reset.**

Free Up Disk Space

The Disk Cleanup utility tidies up your hard drive by deleting old temporary files, installation files, cached copies of web content, and the contents of your Recycle Bin. If you find you are running out of hard disk space, this utility can be your friend, helping you to reclaim some wasted space.

To free up disk space with Disk Cleanup, follow these steps:

1. **Open File Explorer and navigate to This PC by clicking it in the navigation bar on the left.**

2. **Right-click your hard drive (C:) and choose Properties.**

   The Properties dialog box opens.

3. **On the General tab, click Disk Cleanup.**

   The Disk Cleanup utility opens. See Figure 19-9.

FIGURE 19-9

4. **Click Clean Up System Files.**

The utility briefly closes and then reopens, this time including system files that could be deleted.

5. **Click the check boxes for the categories of content you want to delete.**

When a category is selected, information about it appears in the lower part of the dialog box. Some of the categories, when selected, display a View Files button; click this to see the exact files that will be deleted.

6. **Click OK.**

A confirmation box appears.

7. **Click Delete Files.**

It takes a few minutes for the cleanup to complete.

WHAT ABOUT DISK MAINTENANCE TOOLS?

A decade ago, this chapter would have included robust sections that explained how to check a disk for errors and how to defragment its file system. Windows still contains utilities for doing those things, but much of the need for them has gradually gone away due to a combination of better hardware, better Windows error-detection abilities, better Windows automatic scheduling of maintenance tasks, and technology changes.

You can access these utilities from File Explorer. Right-click your hard drive and choose Properties, and then look on the Tools tab for them.

Disk checking for errors was an important task to complete on a regular basis in the early days of computing because old hard drives tended to develop errors rather frequently — both physical bad spots on the disk and errors in the recording of what files were stored on what spots. Nowadays, disks are much more reliable, and Windows also monitors the disk usage closely and lets you know if there are any errors detected. So, running the Error Checking utility is not likely to turn up any problems on a modern system. Don't take my word for it, though — try it if you like.

Similarly, defragmenting a disk was an important task because it improved disk access time — that is, the speed at which the disk could read files into memory. Users don't have to manually trigger the defragmenting (optimizing) utility anymore because Windows schedules it to happen silently and automatically at certain intervals. It's also becoming a moot point because solid-state drives (the newest hard drive technology) don't need defragmenting.

Index

A

accent color, customizing, 165–167

accessibility, Windows
  Eye control feature, 174
  Filter Keys feature, 172
  High Contrast feature, 173
  Magnifier feature, 171
  Mouse Keys feature, 173
  Narrator feature, 171
  On-Screen Keyboard feature, 172
  overview, 170
  Speech Recognition feature, 171
  Sticky Keys feature, 172
  Text Cursor feature, 174
  Text Size feature, 173
  Toggle Keys feature, 172

Account Protection category, Windows Security app, 368

Action Center, 80

active windows, 95

activity history, Windows, 39

Add 3D effects option, Photos app, 320

Add animated text option, Photos app, 320

Add Files button, Skype, 281

adding
  contacts to Microsoft Teams, 277–278
  files to OneDrive, 259–261

items to Favorites list, 200–201

user accounts, 54–56

addresses, Mail app, 245–247

administrator user account, 57–58

adware, 215, 362

Alarms & Clock app, 116–119

all-in-one desktops, 21–22

Alt+Tab keyboard shortcut, 95

Android operating system, 359

anti-malware protection, 215–216, 235

App & Browser Control category, Windows Security app, 369

applications, 18
  advertising, 102
  Alarms & Clock app, 116–119
  Calculator app, 108–110
  Calendar app, 124
  Camera app, 124
  cloud-based
    Office Online, 256–257
    OneDrive storage, 257–268
    overview, 253–255
  Cortana app, 121–123
  desktop apps, 86–90
  exiting, 85–86
  File Explorer app, 124
  free trials, 102

Get Help app, 124

in-app purchases, 102

installing, 101–104

Mail app, 124
  account settings, 248–251
  addresses, 245–247
  attachments, 247–248
  Folders pane, 238
  forwarding messages, 242–243
  Inbox, 238
  Interface, 238–239
  Message pane, 238–239
  Overview, 236–238
  Personalization option, 251
  receiving messages, 239–241
  replying to messages, 241–242
  sending messages, 243–245

Maps app, 124

Microsoft Store apps, 90–91

Microsoft Windows, 91–94

Movies & TV app, 124, 314–317

moving and copying data between, 97–101

Notepad app, 114–116

OneNote app, 124

overview, 77–81, 107

Paint 3D app, 124

Paint app, 124

People app, 124

applications *(continued)*

Photos app, 124

removing, 104–106, 377–378

repairing, 377–378

shutting down unresponsive apps, 374–376

Skype app, 124

Snip & Sketch app, 125

starting, 81–85

Sticky Notes app, 125

switching between, 95–96

Teams app, 125

troubleshooting, 376–377

Video Editor app, 125

Voice Recorder app, 125

Weather app, 119–121

Windows Fax & Scan app, 125

Windows Media Player app, 125

WordPad app, 110–114

Attach files button, Microsoft Teams, 279

attachments, Mail app, 247–248

audio

music

burning CDs, 338–339

digital music, 326–330

downloading, 339–340

ripping CDs, 337–338

subscribing to streaming service, 340

Windows Media Player, 330–336

Voice Recorder app, 311–313

avatars, 275

B

background, customizing, 164–165

backing up files, 150–151

BigFish Games, 302

binary system, 14

Bing search engine, 197–198

bits, 12, 14

blocking unwanted apps, 220–221

blogs, 270–273, 284

Bluetooth, 183

connecting devices to PC, 357–358

headphones, 326–327

board games, 295

Boggle game, 293

Boggle with Friends game, 301–302

broadband Internet connection, 183–184

broadband modem, 37

browsers

apps, 190

defined, 179

Internet Explorer, 179, 190

Microsoft Edge

collections feature, 207–208

customizing settings, 211

Favorites list, 200–204

History list, 204–205

New Tab page, 208–210

overview, 192–196

pinning tabs, 200

printing web page, 205–206

search engines, 196–198

searching on web page, 199

Mozilla Firefox, 190–191

Opera, 190

overview, 191–192

Safari, 190–191

burning CDs, 338–339

buying a computer

connecting to Internet

Internet Service Provider, 29

wired connection, 28

wireless networking, 28

displays

display adapter, 25

display screen, 24–26

DVD drive, 27

guidelines for, 30–31

hardware

memory, 12–13

motherboard, 12

overview, 11–12

processing, 12

storage, 13–14

input devices

keyboard, 15

pointing devices, 15–16

output devices

displays, 16–17

printers, 17

speakers, 17

overview, 7–8

personal computers

all-in-one, 21–22

desktops, 19–22

laptops, 19–20, 22–23

ports, 24

tablets, 20
tower desktop, 20
reasons to, 8–11
retail stores vs. online, 30
software, 17–18
storage
  capacity, 27
  hard disk drives, 26–27
  solid-state drives, 27
bytes, 12

C

cable Internet connection, 181
Calculator app, 108–110
Calendar app, 124
Camera app, 124, 306–310
card games, 295
casino games, 295
casual games, 293
CDs, 14
cellular Internet
  connection, 182
central processing unit
  (CPU), 12
Chat icon, 80
chat rooms, 273–275
Clipboard
  cut/copy/paste commands,
    97–101
  using with File Explorer, 143
cloud-based applications
  Office Online, 256–257
  OneDrive, 128, 131
    accessing, 257–259
    adding files to, 259–261
    adjusting settings, 265–266

creating new folder, 263–264
Personal Vault folder, 264
sharing files, 261–263
Sync feature, 266–268
overview, 253–255
Collections feature, Microsoft
  Edge, 207–208
Color/Black & White settings,
  printer, 70
Command bar, Windows
  Media Player, 331
Compatibility Mode,
  Windows, 378–379
compressed files, 148–149
connecting to Internet, 28–29,
  180–184
  broadband, 183–184
  cable, 181
  cellular, 182
  dialup, 182
  digital subscriber line, 181
  fiber-optic service, 181
  Internet Service Providers,
    29, 180, 234–235
  satellite, 181–182
  servers, 180
  wired connections, 28
  wireless connections
    Bluetooth, 183
    hotspots, 182
    near-field
      communication, 183
    setting up, 187–189
    standards, 28, 183
Content pane, Windows
  Media Player, 331
cookies, 221–223

copying items
  between applications,
    97–101
  File Explorer, 142–144
Cortana app, 80, 121–123
CPU (central processing
  unit), 12
crashes, 373
Create a video with music
  option, Photos app, 320
Credential Manager, 362
Currency Converter
  calculator, 110
Custom layout, Microsoft
  Edge, 208
customizing
  Microsoft Edge settings, 211
  OneDrive settings, 265–266
  Windows
    accent color, 165–167
    accessibility features,
      170–174
    backgrounds, 164–165
    icons, 167–168
    overview, 153–154
    screen resolution and
      scale, 161–163
    themes, 163–164
    widgets, 168–169
    Windows 10 Start Menu,
      157–160
    Windows 10 taskbar,
      160–161
    Windows 11 Start Menu,
      154–156
    Windows 11 taskbar,
      156–157
    Windows security app, 363

D

A Dark Room game, 303

data mining, 270

Date Calculation calculator, 110

default printers, 67–69

defragmenting, 387

deleting items
  from File Explorer, 144–145
  from OneDrive, 261

desktop, Windows, 40

desktop apps, 86–90
  File tab, 89
  menu bar, 86
  Ribbon, 88–89
  ScreenTips, 87
  toolbars, 87

desktops, 19–22
  all-in-one, 21–22
  overview, 19–20
  tower design, 20

Details pane, File Explorer, 141

Device Performance & Health category, Windows Security app, 369

Device Security category, Windows Security app, 369

dialup Internet connection, 182

digital music, 326–330

digital subscriber line (DSL), 181

Discord platform, 270

discussion boards, 270–273

dishNET, 181

Disk Cleanup utility, 385–386

displays
  cost, 26
  display adapter, 25
  display screen, 24–26
  image quality, 25
  overview, 16–17
  resolution, 26, 161–163
  touchscreen, 16, 22, 26, 44

double-clicking, 42

downloading files
  music, 339–340
  photos and videos, 317
  safely, 216–218

drag-and-drop, 43, 101, 143–144

Draw option, Photos app, 320

drivers, printer, 64

Dropbox, 255

DSL (digital subscriber line), 181

DVD drive, 27

DVDs, 14

dynamic storage, 14

E

Ease of Access settings, Windows
  Eye control, 174
  Filter Keys, 172
  High Contrast, 173
  Magnifier, 171
  Mouse Keys, 173
  Narrator, 171
  On-Screen Keyboard, 172
  overview, 170–171
  Speech Recognition, 171
  Sticky Keys, 172
  Text Cursor, 174
  Text Size, 173
  Toggle Keys, 172

e-commerce, 179

Edge browser, 190
  address bar, 192–193
  Collections feature, 207–208
  customizing settings, 211
  displaying web content, 195
  Favorites list, 200–204
  History list, 204–205
  icons, 194
  navigation tools, 193
  New Tab page, 208–210
  overview, 192–196
  password manager, 232
  pinning tabs, 200
  printing web page, 205–206
  privacy settings, 221–223
  search engines, 196–198
  searching on web page, 199

Edit Favorite dialog box, Microsoft Edge, 202

Edit option, Photos app, 320

Edit with Paint 3D option, Photos app, 320

Elvenar games, 296

email
  anti-malware protection, 235
  apps, 190
  Gmail, 190, 236, 254
  junk mail filtering, 235
  Mail app, 190
    account settings, 248–251
    addresses, 245–247
    attachments, 247–248
    interface, 238–239

Message pane, 241–243

overview, 236–238

receiving messages, 239–241

sending messages, 243–245

Outlook, 190, 254

overview, 233–234

phishing, 228–229

signing up, 234–236

storage, 235

Emoji button, Microsoft Teams, 279

emoticons, defined, 275

Error Checking utility, 387

Ethernet adapter, 184–185

Ethernet port

defined, 28

home network, 344–345

executable files, 217

Exede, 181

exiting applications, 85–86

external drive, backing up files to, 150–151

Eye control feature, Windows, 174

F

Facebook

games, 293, 295

signup process, 286

facial recognition, 371

Family Options category, Windows Security app, 369

family settings, Windows, 58–60

fantasy and adventures games, 298

Favorites bar, Microsoft Edge, 201

Favorites list, Microsoft Edge, 200–204

adding items to, 200–201

removing items from, 202

Favorites section, Weather app, 121

fiber-optic service (FiOS), 181

File Explorer, 124

accessing OneDrive storage from, 257

adding files to OneDrive with, 259–260

backing up files, 150–151

compressed files, 148–149

copying items, 142–144

creating folders, 129–130

creating shortcuts, 146–147

deleting items, 144–145

Details pane, 141

file extensions, 146

moving items, 130, 134–136, 142–144

OneDrive and, 131

overview, 132–133

predefined folders, 129

Preview pane, 141

Quick Access list, 149–150

renaming items, 145–146

Search feature, 136–139

selecting multiple items, 142

subfolders, 130

viewing file listings, 139–142

File Explorer icon, Windows, 80

file management. *See also* File Explorer

overview, 127–128

system files, 131

file sharing

choosing which folders to share, 351–353

file-sharing sites, 255

on home network, 348–350

File tab

desktop apps, 89

Notepad app, 115

WordPad app, 113–114

Filter Keys feature, Windows, 172

financial applications, 255

fingerprint recognition, 371

FiOS (fiber-optic service), 181

Firewall & Network Protection category, Windows Security app, 368

first-party cookies, 221

first-person shooter games, 296

Flickr, 255

Focused layout, Microsoft Edge, 208

folders. *See* File Explorer

Folders pane, Mail app, 238

Format button, Microsoft Teams, 279

Format tab, Notepad app, 116

forwarding messages, Mail app, 242–243

freeware, 299

G

games
  BigFish Games, 302
  board games, 295
  Boggle with Friends game, 301–302
  card games, 295
  casino games, 295
  casual games, 293
  on CD or DVD, 292
  A Dark Room game, 303
  downloading files, 291
  fantasy and adventures, 298
  freeware, 299
  game console systems, 292
  mobile, 292, 299–300
  multi-game website, 292
  number games, 293
  The Paperclip game, 304
  Pinball FX3 game, 302
  public domain software, 299
  Puzzle Baron game, 303
  shareware, 299
  shooter games, 296
  The Sims game, 301
  simulations, 298
  single-game website, 292
  on social media, 292
  sports and driving, 298
  Taonga game, 300, 301
  word games, 293–294
  world building, 296
GB (gigabytes), 12–13
gestures, touchscreen, 44
Get Help app, 124
gigabytes (GB), 12–13

gigahertz, 12
Giphy button, Microsoft Teams, 279
Gmail, 190, 235–236, 254
Google Chrome, 190
Google Docs, 254
Google Meet app, 190
Google search engine, 197
grayscale mode, printer, 70
Groove Music app, 330
GUI (graphical user interface), 18

H

hard disk drives (HDDs), 13–14, 26–27
hardware, 11–14
  defined, 7
  Internet
    Ethernet adapter, 184–185
    modem, 37, 184
    network adapter, 185–186
    router, 185
  memory, 12–13
  motherboard, 12
  overview, 11–12
  processing, 12
  setting up, 34–37
  storage, 13–14
HDDs (hard disk drives), 13–14, 26–27
HDMI (High Definition Multimedia Interface), 34–35
Hello feature, Windows, 39
High Contrast feature, Windows, 173

Historical Weather section, Weather app, 121
History list, Microsoft Edge, 204–205
home network. *See* networks
Home tab, WordPad app, 111
hotspots, 358–359
HughesNet, 181
hyperlinks
  anti-malware protection, 235
  defined, 180
  Windows 10 taskbar, 161

I

icons
  customizing, 167–168
  defined, 79
  unpinning, 45
IE (Internet Explorer), 179, 190
IM (instant messaging), 19
  avatars, 275
  emoticons, 275
  shortcut text, 276
  stickers, 275
Inbox, Mail app, 238
information exposure
  clubs and organizations, 224
  employers, 223–224
  family members and friends, 224
  government agencies, 224
  newspapers, 224
  online directories, 224–225
Informational layout, Microsoft Edge, 208
inkjet printers, 62–63

InPrivate browsing, 219

input devices
  keyboard, 15
  pointing devices
    mouse, 15, 42–43
    touchpad, 16
    touchscreen, 16, 22, 26, 44
    trackball, 15

Inspirational layout, Microsoft
  Edge, 208

installing
  camera software, 318
  Microsoft Store apps, 90–91
  printers, 64
  Windows apps, 101–104

instant messaging (IM), 19
  avatars, 275
  emoticons, 275
  shortcut text, 276
  stickers, 275

Internet. *See also*
  Microsoft Edge
  browsers
    apps, 190
    defined, 179
    Internet Explorer, 179, 190
    Mozilla Firefox, 190–191
    Opera, 190
    overview, 191–192
    Safari, 190–191
  connections
    automatically
      connecting, 38
    Bluetooth, 183
    broadband, 183–184
    cable, 181

cellular, 182
dialup, 182
digital subscriber line, 181
fiber-optic service, 181
Internet Service
  Provider, 180
near-field
  communication, 183
satellite, 181–182
servers, 180
Wi-Fi, 182, 183, 187–189
defined, 178
e-commerce, 179
hardware
  Ethernet adapter, 184–185
  modem, 37, 184
  network adapter, 185–186
  router, 185
hyperlinks, 180
overview, 177
software
  browser app, 190
  email app, 190
  overview, 189
  video-calling app, 190
uniform resource
  locator, 179
web pages, 179
websites, 179
World Wide Web, 179
Internet Explorer (IE), 179, 190
Internet Service Provider
  (ISP), 29
  defined, 180
  providing email, 234–235
iOS, 359

J
junk mail filtering, 235

K
keyboard
  Alt+Tab keyboard
    shortcut, 95
  defined, 15
  On-Screen Keyboard, 172
  Windows keyboard
    layout, 38
keyword searches, 198

L
language, Windows, 38
laptops, 19–20, 22–23
laser printers, 62–63
license agreement,
  Windows, 38
links
  anti-malware protection, 235
  defined, 180
  Windows 10 taskbar, 161
List pane, Windows Media
  Player, 332
locking computer, 49–50

M
Magnifier feature,
  Windows, 171
Mail app, 124, 190
  account settings, 248–251
  addresses, 245–247
  attachments, 247–248
  Folders pane, 238

Mail app *(continued)*
  forwarding messages, 242–243
  Inbox, 238
  interface, 238–239
  Message pane, 238–239
  overview, 236–238
  Personalization option, 251
  receiving messages, 239–241
  replying to messages, 241–242
  sending messages, 243–245
malware, 362
  adware, 215
  anti-malware utility, 215–216
  spyware, 215
  viruses, 215
Maps app, 124
Maps section, Weather app, 120
memory, 12–13
menu bar, desktop apps, 86
Message pane, Mail app, 238–239
messages, email
  forwarding, 242–243
  receiving, 239–241
  replying to, 241–242
  sending, 243–245
MFDs (multifunction devices), 63
microphones
  built-in, 306
  external USB, 306
  troubleshooting, 311

Microsoft account
  changing password, 369–370
  creating, 38–39
Microsoft Edge, 190
  address bar, 192–193
  Collections feature, 207–208
  customizing settings, 211
  displaying web content, 195
  Favorites list, 200–204
  History list, 204–205
  icons, 194
  navigation tools, 193
  New Tab page, 208–210
  overview, 192–196
  password manager, 232
  pinning tabs, 200
  printing web page, 205–206
  privacy settings, 221–223
  search engines, 196–198
  searching on web page, 199
Microsoft Edge icon, 80
Microsoft OneDrive, 255
  accessing, 257–259
  adding files to, 259–261
  creating new folder, 263–264
  customizing settings, 265–266
  deleting files from, 261
  File Explorer and, 128, 131
  mirror, 257
  Personal Vault folder, 264
  sharing files from, 261–263
  Sync feature, 266–268
Microsoft Store apps, 90–91
Microsoft Teams, 125, 270

adding contacts, 277–278
Attach files button, 279
Emoji button, 279
Format button, 279
Giphy button, 279
Microsoft Windows. *See also* applications
  Action Center, 80
  active window, 95
  activity history, 39
  administrator user account, 57–58
  changing Microsoft account password, 369–370
  changing sign-in, 371–372
  Chat icon, 80
  Compatibility Mode, 378–379
  control buttons, 94
  Cortana icon, 80
  creating more user accounts, 54–56
  customizing
    accent color, 165–167
    accessibility features, 170–174
    backgrounds, 164–165
    icons, 167–168
    overview, 153–154
    screen resolution and scale, 161–163
    themes, 163–164
    widgets, 168–169
    Windows 10 Start Menu, 157–160
    Windows 10 taskbar, 160–161

Windows 11 Start Menu, 154–156

Windows 11 taskbar, 156–157

defragmenting, 387

desktop, 40

Disk Cleanup utility, 385–386

Error Checking utility, 387

family settings, 58–60

File Explorer icon, 80

Hello feature, 39

Home edition, 19

icons, 79

Internet connection, 38

keyboard layout, 38

language, 38

license agreement, 38

locking computer, 49–50

maximized, 92

Microsoft account, 38–39

Microsoft Edge icon, 80

minimized, 92–93

notification area, 41–42

Notification area, 80

overview, 91–92

PIN, 39

privacy settings, 39

region, 38

removing apps, 377–378

repairing apps, 377–378

resetting, 383–385

restarting, 51–52

restored, 92–94

Search feature, 79

security

changing Microsoft account password, 369–370

changing sign-in, 371–372

Credential Manager, 362

overview, 361–362

software, 362–364

Windows Defender Firewall, 362

Windows Security app, 367–369

Windows Update tool, 364–367

setting up, 38–42

setting up printer to work with, 64–67

Settings icon, 80

shutting down unresponsive apps, 374–376

signing in, 49, 371–372

signing out, 48–49

Sleep mode, 52–53

standard user account, 57

Start button, 41–42, 79

Start menu

overview, 44

Pinned section, 45, 47

Recommended section, 45

shortcuts, 46–47

User icon, 46

switching accounts, 49–50

System Protection feature, 380

System Restore, 379–382

Task View, 80

taskbar, 40–41

title bar, 94

troubleshooting, 382–383

troubleshooting applications, 376–377

versions, 19

Widgets icon, 80

Mint service, 255

mobile devices

mobile apps, 90, 292, 299–300

operating systems, 17

modem, 37, 184

monitors

cost, 26

display adapter, 25

display screen, 24–26

image quality, 25

overview, 16–17

resolution, 26, 161–163

touchscreen, 16, 22, 26, 44

Morningstar Portfolio Manager, 255

motherboard, 12

mouse

clicking, 42

defined, 15

double-clicking, 42

dragging, 43

right-clicking, 43

scrolling, 43

Mouse Keys feature, Windows, 173

Movies & TV app, 124, 314–317

moving items

between applications, 97–101

File Explorer, 130, 134–136, 142–144

Mozilla Firefox, 190–191

multifunction devices (MFDs), 63

multi-game website, 292

music
  burning CDs, 338–339
  digital music, 326–330
  downloading, 339–340
  ripping CDs, 337–338
  subscribing to streaming service, 340
  Windows Media Player, 330–336

N

Narrator feature, Windows, 171

Navigation pane, Windows Media Player, 331

near-field communication (NFC), 183

network adapter, 185–186

networks
  connecting Bluetooth devices, 357–358
  enabling wireless router security, 346–348
  file sharing, 348–354
  hotspots, 358–359
  overview, 343
  setting up, 344–346
  sharing printers, 354–356

New Tab page, Microsoft Edge, 208–210
  Custom layout, 208
  Focused layout, 208
  Informational layout, 208
  Inspirational layout, 208

News section, Weather app, 121

NFC (near-field communication), 183

notebooks, 19–20, 22–23

Notepad app, 114–116

Notification area, 80

notification area, Windows, 41–42

number games, 293

O

Office Online, 256–257

OneDrive, 255
  accessing, 257–259
  adding files to, 259–261
  creating new folder, 263–264
  customizing settings, 265–266
  deleting files from, 261
  File Explorer and, 128, 131
  mirror, 257
  Personal Vault folder, 264
  sharing files from, 261–263
  Sync feature, 266–268

OneNote app, 124

online auctions, 31

online dating, 287–288

online shopping, 30–31

On-Screen Keyboard feature, Windows, 172

Open Richtext Editor button, Skype, 281

Opera browser, 190

operating systems (OS), 8. See also Windows operating system

Android, 359

defined, 1

iOS, 359

optical discs, 14

Outlook, 190, 254

output devices
  displays
    cost, 26
    display adapter, 25
    display screen, 24–26
    image quality, 25
    overview, 16–17
    resolution, 26, 161–163
    touchscreen, 16, 22, 26, 44
  printers
    cost, 62–63
    default, 67–69
    defined, 17, 61
    drivers, 64
    inkjet, 62–63
    installing, 64
    laser, 62–63
    multifunction devices, 63
    photo printing, 63
    Plug and Play process, 64
    print queue, 72–73
    printer sharing, 64
    removing, 73–74
    setting preferences, 69–71
  speakers, 17

P

Paint 3D app, 124

Paint app, 124

Paltalk, 273

Paper Size setting, printers, 71

Paper Source setting, printers, 71

The Paperclip game, 304

passwords
changing Microsoft account password, 369–370
creating strong passwords, 230–232
picture password, 371
wireless routers, 346

People app, 124

peripherals, 24

personal computers (PCs)
desktops
all-in-one, 21–22
overview, 19–20
tower design, 20
laptops, 19–20, 22–23
ports, 24
tablets, 20

Personal Vault folder, OneDrive, 264

phishing, 228–229

phones. *See* mobile devices; smartphones

photos and videos
Camera App, 306–310
downloading, 317
Movies & TV app, 124, 314–317
overview, 305–306
photo printing, 63
Photos app, 124, 319–321
photo-sharing sites, 255
streaming, 317
uploading, 318

Video Editor app, 321–323
Voice Recorder app, 311–313

picture password, 371

PIN, 39, 371

Pinball FX3 game, 302

pinch gesture, touchscreen, 44

Pinned section, Windows Start menu, 45, 47

pinning/unpinning items, 158–159, 200

Pinterest, 285

pixels, 158–159

Playback controls, Windows Media Player, 332

playlists, Windows Media Player, 335–336

Plug and Play (PnP) process, 64

pointing devices
mouse
clicking, 42
defined, 15
double-clicking, 42
dragging, 43
right-clicking, 43
scrolling, 43
touchpad, 16
touchscreen, 16, 22, 26, 44
trackball, 15

ports
defined, 24
Ethernet, 28
USB, 24, 35

Preview pane, File Explorer, 141

printers
cost, 62–63
default, 67–69
defined, 17, 61
drivers, 64
inkjet, 62–63
installing, 64
laser, 62–63
multifunction devices, 63
photo printing, 63
Plug and Play process, 64
print queue, 72–73
printer sharing, 64
removing, 73–74
setting preferences
Color/Black & White settings, 70
grayscale mode, 70
Paper Size setting, 71
Paper Source setting, 71
Quality setting, 71
setting up to work with Windows, 64–67
sharing local printer, 354–356
unpacking, 64

printing
photos from camera, 318
web pages, 205–206

privacy settings, 39, 221–223

private messaging, 288

Program Compatibility Troubleshooter, 379

Protection History category, Windows Security app, 369

public domain software, 299

Puzzle Baron game, 303

Q

Quality setting, printers, 71

Quick Access list, File Explorer, 149–150

Quick Access Toolbar (QAT), 112

R

random access memory (RAM), 12–13

rebooting, 51–52

receiving messages, Mail app, 239–241

Recommended section, Windows Start menu, 45

Record a message button, Skype, 281

region, Windows, 38

removing
  applications, 104–106, 377–378
  items from Favorites list, 202
  printers, 73–74

renaming items, File Explorer, 145–146

replying to messages, Mail app, 241–242

resetting Windows, 383–385

resolution, screen, 26, 161–163

restarting Windows, 51–52

restoring files and folders, 145

Ribbon
  desktop apps, 88–89
  WordPad app

File tab, 113–114
  Home tab, 111
  View tab, 112

right-clicking, 43

ripping CDs, 337–338

RJ-45 jack
  defined, 28
  home network, 344–345

routers, 185

S

Safari browser, 190–191

satellite Internet connection, 181–182

scale, customizing, 161–163

Scrabble game, 293

ScreenTips, desktop apps, 87

scroll bar, 81

scrolling, 43

search engines, 196–198

Search feature, 79, 81
  File Explorer, 136–139
  searching on web page, 199

security
  blocking unwanted apps, 220–221
  downloading files, 216–218
  executable files, 217
  firewalls, 216
  information exposure, 223–228
  InPrivate browsing, 219
  overview, 213–214
  passwords, 230–232
  phishing, 228–229
  privacy settings, 221–223
  security key, 371

SmartScreen Filter, 220

technology risks, 214–216

Windows
  changing Microsoft account password, 369–370
  changing sign-in, 371–372
  Credential Manager, 362
  overview, 361–362
  software, 362–364
  Windows Defender Firewall, 362
  Windows Security app, 367–369
  Windows Update tool, 216, 364–367
  wireless routers, 346–348

selecting multiple items, File Explorer, 142

Send Contacts to This Chat button, Skype, 281

sending messages, Mail app, 243–245

servers, Internet, 180

service set IDs (SSIDs), 188

setting up computer. *See also* Microsoft Windows
  hardware, 34–37
  mouse
    clicking, 42
    double-clicking, 42
    dragging, 43
    right-clicking, 43
    scrolling, 43
  overview, 33–34
  shutting down computer, 53–54
  touchscreen, 44

Settings app, 377–378

Settings icon, 80

shareware, 299

sharing photos, 318

shooter games, 296

shortcut text, 276

shortcuts
  File Explorer, 146–147
  Windows Start menu, 46–47

shutting down computer, 53–54

sign-in methods, Windows, 49
  changing, 371–372
  facial recognition, 371
  fingerprint recognition, 371
  picture password, 371
  PIN, 371
  security key, 371

signing out, Windows, 48–49

signing up
  email, 234–236
  social networking, 285–287

The Sims game, 301

simulations, game, 298

single-game website, 292

Skype, 8, 124, 190, 280–282

Sleep mode, Windows, 52–53

smartphones. *See also* mobile devices
  sharing photos from, 318
  using as hotspot, 358–359

SmartScreen Filter, 220

Snip & Sketch app, 125

social engineering, 270

social journaling sites, 284–285

social networking
  blogs, 270–273, 284
  chat rooms, 273–275

discussion boards, 270–273

games, 292

instant messaging, 275–276

Microsoft Teams, 277–279

online dating, 287–288

overview, 269–270

signing up for, 285–287

Skype, 280–282

social journaling sites, 284–285

webcams, 282–283

wikis, 284

software
  applications, 18
  camera, 318
  defined, 7
  graphical user interface, 18
  Internet
    browser app, 190
    email app, 190
    overview, 189
    video-calling app, 190
  operating system, 17–18
  utilities, 18
  Windows, 19, 362–364

solid-state drives (SSDs), 13–14, 27

solid-state storage, 14

speakers, 17

Speech Recognition feature, Windows, 171

spooled print jobs, 72

sports and driving games, 298

spyware, 215, 362

SSDs (solid-state drives), 13–14, 27

SSIDs (service set IDs), 188

standard user account, 57

Start button, Windows, 41–42, 79, 81

Start menu, Windows
  overview, 44
  Pinned section, 45, 47
  Recommended section, 45
  shortcuts, 46–47
  User icon, 46
  Windows 10, 157–160
  Windows 11, 154–156

static storage, 14

stickers, 275

Sticky Keys feature, Windows, 172

Sticky Notes app, 125

storage, 13–14, 26–28
  capacity, 27
  hard disk drives, 26–27
  Microsoft OneDrive, 255
    accessing, 257–259
    adding files to, 259–261
    creating new folder, 263–264
    customizing settings, 265–266
    deleting files from, 261
    File Explorer and, 128, 131
    mirror, 257
    Personal Vault folder, 264
    sharing files from, 261–263
    Sync feature, 266–268
  solid-state drives, 27

Storage Sense feature, 217

streaming services, 340

subfolders. *See* File Explorer

Sudoku game, 293

swipe gesture, touchscreen, 44

Sync feature, OneDrive, 266–268

system files, 131

System Protection feature, 380

System Restore, Windows, 379–382

T

tablets, 20. *See also* mobile devices

Taonga game, 300, 301

Task Manager window, 374

Task View, 80, 95

taskbar, Windows

  customizing

    Windows 10, 160–161

    Windows 11, 156–157

  overview, 40–41

  using to switch between apps, 95

technology risks

  malware

    adware, 215

    anti-malware utility, 215–216

    spyware, 215

    viruses, 215

  overview, 214–215

tethering, 358

Text Cursor feature, Windows, 174

Text Size feature, Windows, 173

themes, customizing, 163–164

third-parties cookies, 221

third-person shooter games, 296

Toggle Keys feature, Windows, 172

toolbars, desktop apps, 87

touchpad, 16

touchscreen, 22, 26

  defined, 16

  pinching, 44

  swiping, 44

  unpinching, 44

tower desktops, 20

trackball, 15

troubleshooting

  applications, 376–377

  microphones, 311

  printing problem, 72

  Windows, 382–383

TrueKey, 230

tweeting, 284

Twitter, 284

two-in-one laptops, 22–23

U

uniform resource locator (URL), 179

Universal Serial Bus (USB) ports, 24, 35

Uno game, 295

unpinch gesture, touchscreen, 44

updating Windows, 364–367

URL (uniform resource locator), 179

USB (Universal Serial Bus) ports, 24, 35

USB flash drive, backing up files to, 150–151

User Access Control window, 217

user accounts

  adding additional user accounts, 54–56

  administrator, 57–58

  standard, 57

  switching, 49–50

User icon, Windows Start menu, 46

utilities, 18

V

Video Editor app, 125, 321–323

video-calling, 190

videos. *See* photos and videos

View tab, WordPad app, 112

Viewbook, 255

Virus & Threat Protection category, Windows Security app, 368

viruses, 215, 362

Voice Recorder app, 125, 311–313

Volume Converter calculator, 110

W

Weather app

  Favorites section, 121

  Historical Weather section, 121

  Maps section, 120

  News section, 121

web (World Wide Web), defined, 179

web browsing

Microsoft Edge
  collections feature,
    207–208
  customizing settings, 211
  Favorites list, 200–204
  History list, 204–205
  New Tab page, 208–210
  overview, 192–196
  pinning tabs, 200
  printing web page, 205–206
  search engines, 196–198
  searching on web
    page, 199
  overview, 191–192
web pages
  defined, 179
  printing, 205–206
  searching on, 199
webcams, 282–283, 306–308
websites, defined, 179
widgets
  customizing, 168–169
  defined, 80
  Widgets icon, 80
Wi-Fi Internet connection
  hotspots, 182
  setting up, 187–189
  standards, 28
  wireless networking vs., 183
wikis, 284
Windows Defender
  Firewall, 362
Windows Fax & Scan app, 125
Windows Media Player
  creating playlists, 335–336
  overview, 330–332
  playing music, 332–335

Windows Media Player
  app, 125
Windows operating system.
  See also applications
  Action Center, 80
  active window, 95
  activity history, 39
  administrator user account,
    57–58
  changing Microsoft account
    password, 369–370
  changing sign-in, 371–372
  Chat icon, 80
  Compatibility Mode,
    378–379
  control buttons, 94
  Cortana icon, 80
  creating more user
    accounts, 54–56
  customizing
    accent color, 165–167
    accessibility features,
      170–174
    backgrounds, 164–165
    icons, 167–168
    overview, 153–154
    screen resolution and
      scale, 161–163
    themes, 163–164
    widgets, 168–169
    Windows 10 Start Menu,
      157–160
    Windows 10 taskbar,
      160–161
    Windows 11 Start Menu,
      154–156
    Windows 11 taskbar,
      156–157
  defragmenting, 387

  desktop, 40
  Disk Cleanup utility,
    385–386
  Error Checking utility, 387
  family settings, 58–60
  File Explorer icon, 80
  Hello feature, 39
  Home edition, 19
  icons, 79
  Internet connection, 38
  keyboard layout, 38
  language, 38
  license agreement, 38
  locking computer, 49–50
  maximized, 92
  Microsoft account, 38–39
  Microsoft Edge icon, 80
  minimized, 92–93
  notification area, 41–42
  Notification area, 80
  overview, 91–92
  PIN, 39
  privacy settings, 39
  region, 38
  removing apps, 377–378
  repairing apps, 377–378
  resetting, 383–385
  restarting, 51–52
  restored, 92–94
  Search feature, 79
  security
    changing Microsoft account
      password, 369–370
    changing sign-in, 371–372
    Credential Manager, 362
    overview, 361–362
    software, 362–364

Windows operating
system *(continued)*
 Windows Defender
  Firewall, 362
 Windows Security app,
  367–369
 Windows Update tool,
  364–367
 setting up, 38–42
 setting up printer to work
  with, 64–67
 Settings icon, 80
 shutting down unresponsive
  apps, 374–376
 signing in, 49, 371–372
 signing out, 48–49
 Sleep mode, 52–53
 standard user account, 57
 Start button, 41–42, 79

Start menu
 overview, 44
 Pinned section, 45, 47
 Recommended section, 45
 shortcuts, 46–47
 User icon, 46
 switching accounts, 49–50
 System Protection
  feature, 380
 System Restore, 379–382
 Task View, 80
 taskbar, 40–41
 title bar, 94
 troubleshooting, 376–377,
  382–383
 versions, 19
 Widgets icon, 80
wired connections, 28

wireless connections
 Bluetooth, 183
 hotspots, 182
 near-field
  communication, 183
 setting up, 187–189
 standards, 183
wireless routers, 346–348
word games, 293–294
Word Online, 111
WordPad app, 110–114
Words with Friends game,
 293–294
world building games, 296
World Wide Web (web), 179

Z

Zoom app, 8, 190, 270

About the Author

Faithe Wempen, M.A., has been educating people about computers for nearly 40 years. She is a Microsoft Office Master Instructor, a CompTIA A+ certified PC repair technician, and the author of over 160 books on computer hardware and software, including *Office 2021 For Seniors for Dummies* and *Outlook 2021 for Dummies*. She also teaches Computer Hardware and Software Architectures at Indiana University/Purdue University at Indianapolis and designs custom online technology courses for several leading online schools. In her spare time, she's a fitness enthusiast, and you'll find her at the gym nearly every day having fun and staying healthy with her fellow seniors.

Dedication

To Margaret, who has been making it possible all these years.

Author's Acknowledgments

A book like this represents the work of many people. Without the contributions of all these team members, this book would not be the accurate, professional product you are holding in your hands right now. Thanks to the entire Wiley editorial and production team for another job well-done.

Acquisitions editor Greg Tubach set up the project initially and got the ball rolling for the team. Christopher Morris kept us all on track and on schedule and checked my grammar and spelling and pointed out where we could do better. Technical editor Kirk Kleinschmidt made sure that everything I was telling you was accurate.

Publisher's Acknowledgments

Acquisitions Editor: Greg Tubach

Project Editor: Christopher Morris

Copy Editor: Christopher Morris

Technical Editor: Kirk Kleinschmidt

Production Editor: Tamilmani Varadharaj

Cover Image: © Jose Luis Pelaez Inc/ Getty Images